Synthesis Lectures on Data Mining and Knowledge Discovery

Series Editors

Jiawei Han, University of Illinois at Urbana-Champaign, Urbana, IL, USA

Lise Getoor, University of California, Santa Cruz, USA

Johannes Gehrke, Microsoft Corporation, Redmond, WA, USA

The series focuses on topics pertaining to data mining, web mining, text mining, and knowledge discovery, including tutorials and case studies. Potential topics include: data mining algorithms, innovative data mining applications, data mining systems, mining text, web and semi-structured data, high performance and parallel/distributed data mining, data mining standards, data mining and knowledge discovery framework and process, data mining foundations, mining data streams and sensor data, mining multi-media data, mining social networks and graph data, mining spatial and temporal data, pre-processing and post-processing in data mining, robust and scalable statistical methods, security, privacy, and adversarial data mining, visual data mining, visual analytics, and data visualization.

Chuan Shi · Xiao Wang · Cheng Yang

Advances in Graph Neural Networks

 Springer

Chuan Shi
Beijing University of Posts
and Telecommunications
Beijing, China

Xiao Wang
Beijing University of Posts
and Telecommunications
Beijing, China

Cheng Yang
Beijing University of Posts
and Telecommunications
Beijing, China

ISSN 2151-0067 ISSN 2151-0075 (electronic)
Synthesis Lectures on Data Mining and Knowledge Discovery
ISBN 978-3-031-16176-6 ISBN 978-3-031-16174-2 (eBook)
https://doi.org/10.1007/978-3-031-16174-2

This Springer imprint is published by the registered company Springer Nature Switzerland AG
The registered company address is: Gewerbestrasse 11, 6330 Cham, Switzerland

Foreword

Relational structures are ubiquitous in the real world, such as social relations between people, transaction relations between companies, and biological relations between proteins. Graphs and networks are the most common way to characterize these structured data, where objects/relations are projected into nodes and edges, respectively. With the great success of machine learning and deep learning techniques, how to numerically represent a graph has become a fundamental problem in network analysis. In particular, graph representation learning, which aims to encode each node in a network into a low-dimensional vector, has attracted much attention during the last decade. More recently, representation learning methods based on graph neural networks (GNNs) show their superiority on various graph-based applications, and become the state-of-the-art paradigm for graph representation learning. GNNs work well for both node-level and graph-level tasks, and immensely contribute to the depth and breadth of the adoption of graph representation learning in real-world applications: ranging from classical graph-based applications such as recommender systems and social network analysis, to new frontiers such as combinational optimization, physics, and health care. The wide applications of GNNs enable diverse contributions and perspectives from disparate disciplines and make this research field truly interdisciplinary.

This book provides a comprehensive introduction to the foundations and frontiers of graph neural networks. It mainly consists of three parts: Fundamental Definitions and Development of GNNs in Part I (Chaps. 1 and 2); Frontier topics about GNNs in Part II (Chaps. 3 and 7), and Future Directions for GNNs in Part III (Chap. 8). The book starts from the basics of graph representation learning, and extensively introduces the cutting-edge research directions of GNNs, including heterogeneous GNNs, dynamic GNNs, hyperbolic GNNs, distilling GNNs, etc. The basic knowledge can help readers quickly understand the merits of GNNs, while the various topics of advanced GNNs are expected to inspire readers to develop their own models. Both beginners and experienced researchers from academia or industry are believed to benefit from the content of this book.

The authors of this book have worked on graph representation learning for years, and developed a series of fundamental algorithms. As my first visiting scholar from mainland China, Chuan Shi and I have built a close collaboration since 2010. He has done great work in heterogeneous information network analysis and promoted the development of this field. Xiao Wang and Cheng Yang, who published several top-cited papers in graph representation learning, are rising-star scholars in network analysis community. I know these excellent young researchers built a fast-rising laboratory focusing on Graph dAta Mining and MAchine learning (named GAMMA Lab, directed by Chuan Shi). This book systematically summarizes contributions of GAMMA Lab in the domain of graph neural networks. I hope you can learn from this book and enjoy it.

Philip S. Yu
UIC Distinguished Professor and Wexler
Chair in Information Technology
University of Illinois Chicago
Chicago, USA

Preface

In the era of big data, graph data has attracted considerable attention, ranging from social networks, biological networks to recommendation systems. For example, in social network, the user and their behavior can be modeled as graph; in chemistry, the molecular structure is naturally a graph; and in text analysis, the relations among words, sentences, and texts can also be modeled as a graph. Despite data may be generated from different fields with various modalities, they all can be considered as a graph, implying that graph will make a profound impact on every walk of life. Naturally, graph analysis is of great scientific and application values.

To bridge the gap between graph data with the real-world applications, one fundamental problem is the graph representation learning, i.e., how to learn the low-dimensional vector for nodes in a graph, so that the applications can be conducted on the new learned vectors instead of original graph structures. Deep learning, which has already well demonstrated their ability on other fields, e.g., computer vison, has also become a promising technique to deal with graph data. Different from previous graph representation learning which mainly focuses on preserving topology, graph neural networks learn the node representation by propagating node features along topology in a layer-by-layer manner. In this way, the learned representation naturally encodes the effective information from both of node feature and topology. To date, graph neural network has become a typical neural network in deep learning. Not surprisingly, we have witnessed the impressive performance of graph neural networks on various real-world applications, including but not limited to recommender systems and biological field. The increasing number of works on graph neural networks indicates a global trend in both academic and industrial communities. Therefore, there is a pressing demand for comprehensively summarizing and discussing the basic and advanced graph neural networks.

This book serves the interests of specific reader groups. Generally, the book is intended for anyone who wishes to understand the fundamental problems and techniques of graph neural networks. In particular, we hope that university students, researchers, and engineers in universities and IT companies will find this book inspiring. This book is divided into three parts, and the readers are able to quickly understand this field through the first part, deeply study the advanced topics of graph neural networks with the second part, and learn the future directions in the third part.

- In the first part (Chaps. 1–2), we first present an overview of the basic concepts of different graphs, and the development of graph neural networks, including several typical graph neural networks. This part will help readers rapidly understand the overall development of this field. In particular, in Chap. 1, the basic concepts and definitions, as well as the development of graph neural networks, will be summarized. The fundamental graph neural networks, including GCN, etc., will be introduced in Chap. 2.
- In the second part (Chaps. 3–7), we then provide an in-depth and detailed introduction of representative graph neural network techniques. This part will help readers understand the fundamental problems in this field, and illustrate how to design the advanced graph neural networks for these problems. In particular, the homogeneous graph neural networks are discussed in Chap. 3, including the multi-channel graph neural networks, etc. In Chap. 4, the heterogeneous graph neural networks are presented, mainly focusing on the heterogeneous graph propagation network, etc. After that, we introduce the dynamic graph neural networks in Chap. 5, which consider the temporal graph, the dynamic heterogeneous Hawkes process, and temporal heterogeneous graph. Then, in Chap. 6, we introduce the hyperbolic graph neural networks, covering the hyperbolic graph attention networks and Lorentzian graph convolutional neural networks, etc. Finally, the distilling graph neural networks are presented in Chap. 7, including the knowledge distillation of graph neural networks and adversarial knowledge distillation, etc.
- In the third part (Chap. 8), we make the conclusion and discuss the future research directions. Despite a large number of graph neural network methods being proposed, many important open problems are still not well explored, such as the robustness and the fairness of graph neural networks. When graph neural networks are applied to real-world applications, especially some risk-sensitivity areas, these problems need to be carefully considered. We summarize the future research directions here.

Writing a book always involves more people than just the authors. We would like to express our sincere thanks to all those who work with us on this book. In particular, this book is written in collaboration with BUPT GAMMA Lab, which is compiled by Chuan Shi, Xiao Wang, and Cheng Yang. The writers of each chapter include Mengmei Zhang, Meiqi Zhu, Deyu Bo, Ruijia Wang, Houye Ji, Nian Liu, Yugang Ji, Yuanfu Lu, Yiding Zhang, Jiawei Liu, Yuanxin Zhuang, Yuxin Guo, Tianyu Zhao, and Yaoqi Liu. In addition, the work is supported by National Natural Science Foundation of China (Nos. U20B2045, U1936220, 61772082, 61702296, 62002029, 62172052). We also thank the supports of these grants. Finally, we thank our families for their wholehearted support throughout this book.

Beijing, China   Chuan Shi
                                                                                                          Xiao Wang
                                                                                                          Cheng Yang

Contents

Introduction

1.1 Basic Concepts

Graph data is used to describe pairwise relations for real-world data from many different domains, including social science, chemistry, biology, and so on. We will first summarize the basic concepts in graph modeling.

1.1.1 Graph Definitions and Properties

In this section, we focus on the simple unweighted graphs and introduce important definitions used in graphs.

Definition 1.1 (*Graph*) A graph can be denoted as $G = \{\mathcal{V}, \mathcal{E}\}$, where $\mathcal{V} = \{v_1, \ldots, v_{|\mathcal{V}|}\}$ represents the node set and $\mathcal{E} = \{e_1, \ldots, e_{|\mathcal{E}|}\}$ represents the edge set. The edge connecting two nodes can be also represented as (v_1, v_2).

For example, in the social graph, nodes represent people and edges represent social relations like friendship, classmates, teacher-student relationship, or parent-child relationship; in recommendation graph, nodes represent people or items while edges represent the purchasing behaviors; in chemistry, chemical compounds are denoted as graphs with atoms as nodes and chemical bonds as edges. A node v_i is adjacent to another node if and only if there exists an edge between them. A graph $G = \{\mathcal{V}, \mathcal{E}\}$ can be equivalently represented as an adjacency matrix, which describes the connectivity between the nodes.

© The Author(s), under exclusive license to Springer Nature Switzerland AG 2023
C. Shi et al., *Advances in Graph Neural Networks*, Synthesis Lectures
on Data Mining and Knowledge Discovery,
https://doi.org/10.1007/978-3-031-16174-2_1

Definition 1.2 (*Adjacency Matrix*) Given a graph $\mathcal{G}$, we can denote the edge distribution using the adjacency matrix denoted as $\mathbf{A} \in \{0, 1\}^{N \times N}$. The (i, j)-th entry of the adjacency matrix is indicated as $\mathbf{A}_{i,j}$, which represents the connectivity between two nodes v_i and v_j. $\mathbf{A}_{i,j} = 1$ if there exists an edge, otherwise, $\mathbf{A}_{i,j} = 0$.

Especially, in directed graphs, the edge is directed from node to node while in undirected graphs, the order of the two nodes does not make a difference. In an undirected graph, a node v_i is adjacent to v_j, if and only if v_j is adjacent to v_i, thus $\mathbf{A}_{ij} = \mathbf{A}_{ji}$ holds for all v_i and v_j in the graph. Hence, for an undirected graph, its corresponding adjacency matrix is symmetric. Note that without specific mention, we limit our discussion to undirected graphs. With the adjacency matrix, we can easily calculate the number of times that a node is adjacent to other nodes, which is defined as the degree of the node. In graph learning, connected nodes always have effects on similarity modeling. And the connected node sets for node v_i always refer to the neighbors.

Definition 1.3 (*Neighbors*) The neighbor set of a node v_i in the graph $\mathcal{G}$ is denoted as $N(v_i)$, which consists of all nodes that are adjacent to v_i.

Definition 1.4 (*Degree*) The degree of a node v_i can be computed as $d_i = \sum_{j=1}^{N} \mathbf{A}_{i,j}$. The degree of node v_i is equal to the number of nodes in $N(v_i)$, i.e., $d_i = |N(v_i)|$. The diagonal degree matrix can be represented as $\mathbf{D} = diag(d_1, d_2, \ldots, d_n)$.

An illustrative graph with 5 nodes and 7 edges is shown in Fig. 1.1a. In this graph, the set of nodes is represented as $\mathcal{V} = \{v_1, v_2, v_3, v_4, v_5\}$, and the set of edges is $\mathcal{E} = \{e_1, e_2, e_3, e_4, e_5, e_6, e_7\}$. Its adjacency matrix can be denoted as $\mathbf{A}$ in Fig. 1.1b, the one-hop neighbors of node v_2 is the node set $\{v_1, v_3, v_5\}$, the corresponding degree number for node v_2 is 3, and the degree matrix for the hole graph can be denoted as $\mathbf{D}$ in Fig. 1.1c.

In many real-world graphs, there are often features or attributes associated with nodes. This kind of graph-structured data can be treated as graph signals, which capture both the

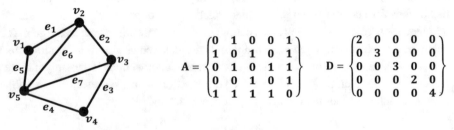

(a) A graph with 5 nodes and 7 edges   (b) The adjacency matrix   (c) The degree matrix

Fig. 1.1 Example graph and the corresponding matrix representations

structure information (or connectivity between nodes) and data (or attributes at nodes). A graph signal aims to map the node feature into real values with a mapping function f defined in the graph domain. Mathematically, the mapping function can be represented as $f : \mathcal{V} \to \mathbb{R}^{N \times b}$, where d is the dimension of the value (vector) associated with each node.

Furthermore, spectral graph theory studies the properties of a graph by analyzing the eigenvalues and eigenvectors of its Laplacian matrix. Then we will define the Laplacian matrix of a graph and then discuss its key properties. Another definition of the Laplacian matrix is a normalized version. Note that the Laplacian matrix is symmetric as both the degree matrix $\mathbf{D}$ and the adjacency matrix $\mathbf{A}$ are symmetric.

Definition 1.5 (*Laplacian Matrix*) For a graph $\mathcal{G}$ with $\mathbf{A}$ as its adjacency matrix, its Laplacian matrix is defined as $\mathbf{L} = \mathbf{D} - \mathbf{A}$, where $\mathbf{D} = diag(d(v_1), \ldots, d(v_N))$ is the diagonal degree matrix.

Definition 1.6 (*Normalized Laplacian Matrix*) For a given graph $\mathcal{G}$ with $\mathbf{A}$ as its adjacency matrix, its normalized Laplacian matrix denoted as $\tilde{\mathbf{L}}$ is

$$\mathbf{L} = \mathbf{D}^{-\frac{1}{2}}(\mathbf{D} - \mathbf{A})\mathbf{D}^{-\frac{1}{2}} = \mathbf{I} - \mathbf{D}^{-\frac{1}{2}}\mathbf{A}\mathbf{D}^{-\frac{1}{2}}. \tag{1.1}$$

1.1.2 Complex Graphs

The above simple graphs we have discussed are all homogeneous. They only have one type of node as well as a single type of edge. Actually, graphs in real-world applications are much more complicated. We further briefly describe some kinds of popular complex graphs.

Firstly, we will talk about the definitions of heterogeneous graph. Heterogeneous graph is also represented as Heterogeneous Information Networks (HIN), which is used to model multiple types of relations between multiple types of nodes in many real-world applications. For example, Fig. 1.2a shows an example of heterogeneous graph, and it can be formally defined as follows:

Definition 1.7 (*Heterogeneous Graph*) A hetereogeneous graph $\mathcal{G}$ consists of a set of nodes $\mathcal{V} = \{v_1, \ldots, v_N\}$ and a set of edges $\mathcal{E} = \{e_1, \ldots, e_M\}$. Each node v and edge e are associated with their type mapping function $\phi_v : \mathcal{V} \to \mathcal{T}_v$ and $\phi_e : \mathcal{E} \to \mathcal{T}_e$, where $|\mathcal{T}_V| + |\mathcal{T}_E| > 2$.

Since a heterogeneous graph contains multiple node types and link types, to understand the whole structure of it, it is necessary to provide a meta-level (or schema-level) description of the graph. Therefore, as the blueprint of heterogeneous, network schema is proposed to give an abstraction of graph. For example, we show an example of network schema in Fig. 1.2b and further give the following definition:

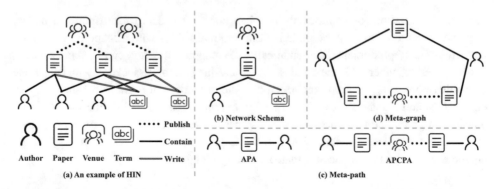

Fig. 1.2 A heterogeneous academic graph, including (**a**) four types of nodes (i.e., Author, Paper, Venue, and Term) and three types of link (i.e., Publish, Contain, and Write), (**b**) network schema, (**c**) meta-path (i.e., Author-Paper-Author and Author-Paper-Conference-Paper-Author), and (**d**) meta-graph

Definition 1.8 (*Network Schema* [248]) Given a heterogeneous graph $\mathcal{HG}$, a network schema $\mathcal{S} = (\mathcal{A}, \mathcal{R})$ can be seen as a meta-template of $\mathcal{G}$ with the node-type mapping function $\phi(v) : \mathcal{V} \to \mathcal{A}$ and the link-type mapping function $\phi(e) : \mathcal{E} \to \mathcal{R}$. Figure 1.2b illustrates the network schema of the academic graph.

Network schema describes the associations between different types of nodes. Based on it, to capture both the structural and semantic correlations in a heterogeneous graph, meta-path (in Fig. 1.2c) or meta-graph (in Fig. 1.2d)-based methods are designed. Meta-path can be used to guide random walks. A meta-path-based random walk is a randomly generated instance of a given meta-path ψ.

Definition 1.9 (*Meta-path* [179]) Given a heterogeneous graph $\mathcal{G}$, a meta-path ψ is denoted as $A_1 \xrightarrow{R_1} A_2 \xrightarrow{R_2} \cdots \xrightarrow{R_l} A_{l+1}$, where $A_i \in \mathcal{T}_n$ and $R_i \in \mathcal{T}_e$ denote certain types of nodes and edges, respectively. The meta-path defines a composite relation between nodes from type A_1 to type A_{l+1} where the relation can be denoted as $R = R_1 \circ R_2 \circ \cdots R_{l-1} \circ R_l$.

Different meta-path captures the semantic relationships from a different view. An example of meta-path is shown in Fig. 1.2, which can be regarded as the combination of meta-paths "APA" and "APCPA", reflecting a high-order similarity of two nodes. For example, the meta-path of "APA" indicates the co-author relationship and "APCPA" represents the co-conference relation. Both of them can be used to formulate the proximity over authors. Note that a meta-graph can be symmetric or asymmetric.

Although meta-path can be used to depict the connections over nodes, it fails to capture the more complex relationship, such as motifs. To address this challenge, the meta-graph is proposed to use a directed acyclic graph of node and link types to capture the more complex relationships between two heterogeneous graph nodes.

Definition 1.10 (*Meta-graph* [88]) A meta-graph $\mathcal{T}$ can be seen as a directed acyclic graph (DAG) composed of multiple meta-paths with common nodes. Formally, meta-graph is defined as $\mathcal{T} = (\mathcal{V}_\mathcal{T}, \mathcal{E}_\mathcal{T})$, where $\mathcal{V}_\mathcal{T}$ is a set of nodes and $\mathcal{E}_\mathcal{T}$ is a set of links. For any node $v \in \mathcal{V}_\mathcal{T}$, $\phi(v) \in \mathcal{A}$; for any link $e \in \mathcal{E}_\mathcal{T}$, $\phi(e) \in \mathcal{R}$.

Furthermore, another kind of graph is widely used to capture interactions between different objects, named bipartite graph. For example, in many e-commerce platforms such as Amazon, the click history of users can be modeled as a bipartite graph where the users and items are the two disjoint node sets, and users' click behaviors form the edges between them. Specifically, we show an example of a bipartite graph in Fig. 1.3a.

Definition 1.11 (*Bipartite Graph*) Given a bipartite graph $\mathcal{G} = \{\mathcal{V}, \mathcal{E}\}$, $\mathcal{V}$ consists of two disjoint node sets $\mathcal{V}_1$ and $\mathcal{V}_2$, i.e., $\mathcal{V} = \mathcal{V}_1 \cup \mathcal{V}_2$ and $\mathcal{V}_1 \cap \mathcal{V}_2 = \emptyset$. Furthermore, there exist no edge for any two nodes from the same node set. For all $e = (v_e^1, v_e^2) \in \mathcal{E}$, we have $v_e^1 \in \mathcal{V}_1$ while $v_e^2 \in \mathcal{V}_2$.

Then we will talk about the graph which can capture temporal information. The graphs mentioned above are all static, where the connections between nodes are fixed when observed. However, in many real-world applications, graphs are constantly evolving as new nodes are added to the graph, and new edges are continuously emerging. For example, we show a graph in Fig. 1.3b whose link information is dynamic. And then we give a formal definition of dynamic graphs.

Definition 1.12 (*Dynamic Graph*) A dynamic graph $\mathcal{G} = \{\mathcal{V}, \mathcal{E}\}$ with node set $\mathcal{V}$ and edge set $\mathcal{E}$ is dynamically changing. Specifically, each node or each edge is associated with timestamp information indicating their emerging timestamps.

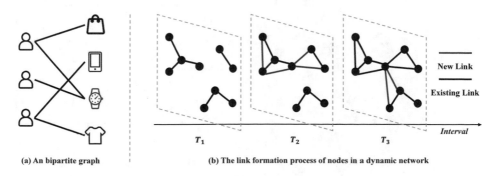

(a) An bipartite graph

(b) The link formation process of nodes in a dynamic network

Fig. 1.3 Example of bipartite graph and dynamic graph

In reality, we may not be able to record all the timestamp of each node and/or each edge. Instead, we always use snapshot to check how the graph evolves, then the observed graph at timestamp t can be represented as G_t. For example, the dynamic graph in Fig. 1.3b also consists of multiple graph snapshots.

1.1.3   Computational Tasks on Graphs

There are a variety of computational tasks proposed for graphs. However, to complete the various tasks on graphs, we first need essential graph representation, i.e., node embedding modeling useful information in the graph. The progress for obtaining graph representation also refers to the graph representation learning.

Definition 1.13 (*Graph Representation Learning* [32]) Graph Representation Learning, also known as network embedding, aims to learn a function $\phi : \mathcal{V} \rightarrow \mathbb{R}^d$ that embeds the nodes $v \in \mathcal{V}$ in a graph into a low-dimensional Euclidean space where $d \ll |\mathcal{V}|$ (Fig. 1.4).

Through graph representation learning, the complex network in the non-Euclidean space is projected into a low-dimensional Euclidean space. Therefore, the high computational cost and low parallelizability issues are well solved.

As for the diverse computational tasks, they can be mainly divided into two categories. One is node-focused tasks, where the entire data is usually represented as one graph with nodes as the data samples. The other is graph-focused tasks, where data often consists of a set of graphs, and each data sample is a graph.

Numerous node-focused tasks have been extensively studied, such as node classification, node ranking, link prediction, and community detection. Next, we mainly discuss two typical

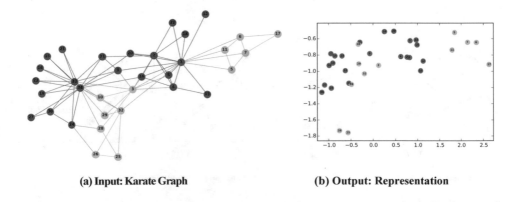

(a) Input: Karate Graph   (b) Output: Representation

Fig. 1.4 A toy example of graph representation. **Left**: the input Karate graph. **Right**: the output node representations. Image is extracted from DeepWalk [154]

tasks, including node classification and link prediction. In many real-world graphs, nodes are associated with useful information, often treated as labels of these nodes. These labels usually help characterize the nodes and can be leveraged for many important applications. However, in reality, it is often difficult to get a full set of labels for all nodes. Hence, we are likely given a graph with only a part of the nodes associated with labels, and we aim to infer the labels for nodes without labels. It motivates the problem of node classification on graphs.

Definition 1.14 (*Node Classification*) In a graph $G = \{\mathcal{V}, \mathcal{E}\}$, some nodes are associated with labels and the node set with these nodes is represented as the labeled set $\mathcal{V}_l \subset \mathcal{V}$. The remaining node set without label information is denoted as the unlabeled set $\mathcal{V}_u = \mathcal{V} - \mathcal{V}_l$ these nodes. Specifically, $\mathcal{V}_u + \mathcal{V}_l = \mathcal{V}$ and $\mathcal{V}_u \cap \mathcal{V}_l = \emptyset$. The goal for node classification task is to predict labels for nodes in $\mathcal{V}_u$, while learning a mapping function ϕ through extracting useful information from G and $\mathcal{V}_l$.

In many real-world applications, graphs are not complete but with missing edges. Some connections exist but are not observed or recorded, which leads to missing edges in the observed graphs. Inferring or predicting these missing edges can benefit many applications.

Definition 1.15 (*Link Prediction*) In a $G = \{\mathcal{V}, \mathcal{E}\}$, $\mathcal{E}$ represents all the observed edges and let $\mathcal{M}$ denote all possible edges between nodes. Then the potential edge set with unobserved edges between nodes can be represented as $\mathcal{E}'$ while $\mathcal{E}' = \mathcal{M} - \mathcal{E}$. The goal of the link prediction task is to predict the edges that are most likely to exist. And a score can be assigned to each of the edges in $\mathcal{E}'$ after link prediction, which indicates how likely the edge exists or will emerge in the future.

Besides the node-focus tasks, there are also numerous graph-focused tasks, such as graph classification, graph matching, and graph generation. Next, we discuss the most representative graph-focused task, i.e., graph classification.

Actually, node classification treats each node in a graph as a data sample and aims to assign labels to these unlabeled nodes. In some applications, each sample can be represented as a graph. For example, in chemoinformatics, chemical molecules can be denoted as graphs where atoms are nodes, and chemical bonds between them are the edges. These chemical molecules may have different properties such as solubility and toxicity, which can be treated as their labels. In reality, we may want to predict these properties for newly discovered chemical molecules automatically. This goal can be achieved by the task of graph classification, which aims to predict the labels for unlabeled graphs. Graph classification always cannot be simply carried out by traditional classification due to the complexity of graph structures. Thus, dedicated efforts are desired. Next, we provide a formal definition of graph classification.

Definition 1.16 (*Graph Classification*) Given a set of labeled graphs $\mathbf{G} = \{(\mathcal{G}_i, y_i)\}$ where y_i is the label of the graph $\mathcal{G}_i$, the goal of the graph classification task is to learn a mapping function which can predict the labels for unlabeled graphs.

1.2 Development of Graph Neural Network

1.2.1 History of Graph Representation Learning

Representation learning on graphs is to learn a set of new node features. It has been greatly developed over the past decades that can be roughly divided into three generations including traditional graph embedding, modern graph embedding, and deep learning on graphs.

As the first generation of graph representation learning, traditional graph embedding has been investigated under the context of spectral clustering, graph-based dimension reduction, and matrix factorization. Then the successful extensions of word2vec to the graph domain have started the second generation of representation learning on graphs, i.e., modern graph embedding. Word2vec is a technique to generate word embeddings, it takes a large corpus of text as input and produces a vector representation for each unique word in the corpus. The huge success of Word2vec in various natural language processing tasks has motivated increasing efforts to apply Word2vec, especially the Skip-gram model to learn node representations. DeepWalk [154] takes the first step to achieve this goal. Specifically, nodes in a given graph are treated as words of an artificial language, and sentences in this language are generated by random walks. Then, it uses the Skip-gram model to learn node representations, which preserves the node co-occurrence in these random walks. Consecutively, several classical graph embedding methods have been proposed as [18, 70, 107, 159, 198].

Given the power and the success of deep neural networks in representation learning, increasing efforts have been made to generalize graph neural networks to graphs. Graph neural network (GNN) is one of the most representative works, which designs a convolutional-based operator in the spatial domain to filter the node attributes by network structures. These methods are known as graph neural networks (GNNs) that can be roughly divided into spatial approaches and spectral approaches. Spectral approaches utilize the spectral view of graphs by taking advantage of graph Fourier transform and the inverse graph Fourier transform. They mainly focus on designing graph spectral filtering operators which can be used to filter certain frequencies of the input signal. Bruna et al. [12] make attempts to design the graph Fourier coefficients while the Chebyshev polynomial filter operator is proposed for [36] to reduce computational cost and make the operator spatially localized. Then [102] simplified Chebyshev filter involving one-hop neighbors. And [213] utilizes a graph wavelet filter to design GNNs. On the other hand, spatial-based graph neural networks have been developed to perform spatial information aggregation involving neighbors. These approaches explicitly leverage the graph structure, such as spatially close neighbors to design spatial filters. Kipf and Welling [102] are the most classical spatial-based GNNs, then the GraphSAGE proposed in [74] first sample neighbors and then introduce a flexible neighbor aggregation strategy with Mean, LSTM, or Pooling aggregators. Then self-attention mechanism [189] is introduced to learn neighborhood edge weights in graph attention networks.

Furthermore, graph neural networks not only work well on node-wise graph analyzing tasks, but also show outstanding performance on graph-focused tasks, such as graph classification. In these tasks, the representation of the whole graph is desired, then numerous pooling methods [60, 126, 225] have been introduced to obtain the whole graph representations.

1.2.2 Frontier of Graph Neural Networks

In the era of deep learning, GNNs have been rapidly developed in the following frontier aspects and corresponding models have been proposed.

Heterogeneous Graph Neural Networks. GNN models have been designed to handle complex graphs such as heterogeneous graphs. In heterogeneous graphs, there are different types of nodes. And to capture both the structural and semantic correlations in the heterogeneous graph, meta-path- or meta-graph-based methods are designed. For example, [25, 244] utilized meta-path to split the heterogeneous graph into several homogeneous graphs. Wang et al. [200] proposed HAN which aggregated information from different types of meta-path-based neighbors through the attention mechanism, and further generated node representations. Hu et al. [83] proposed a heterogeneous graph attention network for short text classification. GTN [230] learned a soft selection of edge types and generated meta-paths automatically, solving the problem of meta-path selection. HGT [84] adopted heterogeneous mutual attention to aggregate meta-relation triplet, and MAGNN [55] leveraged relational rotation encoder to aggregate meta-path instances.

Dynamic Graph Neural Networks. DANE [113] leveraged matrix perturbation theory to capture the changes of adjacency and attribute matrix in an online manner. DynamicTriad [254] imposed the triadic closure process to preserve both structural information and evolution patterns of the dynamic network. CTDNE [139] designed a time-dependent random walk sampling method for learning dynamic network embeddings from continuous-time dynamic networks. HTNE [260] integrated the Hawkes process into network embeddings to capture the influence of historical neighbors on the current neighbors for temporal network embedding. Dyrep [186] utilized a deep temporal point process model to encode structural-temporal information over graphs into low-dimensional representations. To inductively infer embeddings for both new and observed nodes as the graph evolves, [214] proposed the temporal graph attention mechanism based on the classical Bochner's theorem.

Hyperbolic Graph Neural Networks. Recently, node representation learning in hyperbolic spaces has received increasing attention. Nickel and Kiela [140] embedded graphs into hyperbolic spaces to learn the hierarchical node representation. Sala et al. [161] proposed a novel combinatorial embedding approach as well as an approach to multi-dimensional scaling in hyperbolic spaces. To better model hierarchical node representation, [58, 180] embedded the directed acyclic graphs into hyperbolic spaces to learn their hierarchical feature representations. Law et al. [109] analyzed the relation between hierarchical representations and Lorentzian distance. Also, [6] analyzed the hierarchical structure in the multi-relational graph, and embedded them in hyperbolic spaces. Moreover, some researchers began to study deep learning in hyperbolic spaces. Liu et al. [120] generalized deep neural models in hyperbolic spaces, such as recurrent neural networks and GRU. Gülçehre et al. [71] proposed the attention mechanism in hyperbolic spaces. There are some attempts in hyperbolic GCNs recently. Chami et al. [21] proposed graph neural networks in hyperbolic spaces which focuses on the graph classification problem. Chami et al. [21] leveraged hyperbolic graph convolution to learn the node representation in the hyperboloid model. Zhang et al. [242] proposed graph attention network in Poincare ball model to embed some hierarchical and scale-free graphs with low distortion. Bachmann et al. [5] also generalized graph convolutional in a non-Euclidean setting.

Distilling Graph Neural Networks. In fact, there are also a few studies combining knowledge distillation with GNNs. Yang et al. [220], which was proposed in the computer vision area, compressed a deep GCN with large feature maps into a shallow one with fewer parameters using a local structure-preserving module. Reliable Data Distillation (RDD) [240] trained multiple GCN students with the same architecture and then ensembled them for better performance in a manner similar to BAN [57]. Graph Markov Neural Networks (GMNN) [156] can also be viewed as a knowledge distillation method where two GCNs with different reception sizes learn from each other. Note that both teacher and student models in these works are GCNs.

More and more evidence has demonstrated that the third generation of graph representation learning especially graph neural networks (GNNs) has tremendously facilitated computational tasks on graphs including both node-focused and graph-focused tasks. The revolutionary advances brought by GNNs have also immensely contributed to the depth and breadth of the adoption of graph representation learning in real-world applications. For the classical application domains of graph representation learning such as recommender systems and social network analysis, GNNs result in state-of-the-art performance and bring them into new frontiers. Meanwhile, new application domains of GNNs have been continuously emerging such as combinational optimization, physics, and health care. These wide applications of GNNs enable diverse contributions and perspectives from disparate disciplines and make this research field truly interdisciplinary.

1.3   Organization of the Book

This book provides a comprehensive introduction to the foundations and frontiers of graph neural networks. It mainly consists of four parts: Fundamental Definitions and Development of GNNs in Part I (Chaps. 1–2); Frontier topics about GNNs in Part III (Chaps. 3–7), including Homogenous GNNs, Heterogeneous GNNs, Dynamic GNNs, Hyperbolic GNNs, Distilling GNNs, and so on; and Future Direction and Conclusion for GNNs in Part IV (Chap. 8).

- **Part I**: The foundation part in Chapter I focuses on the basics of graphs and the development of graph neural networks. We introduce the key concepts and formally define various types of complex graphs and computational tasks on graphs. Then we discuss the development of graph neural networks. We also mention a variety of frontier graph neural networks. Furthermore, the fundamental part introduces the most representative and basic graph neural network model including graph convolutional networks (GCN), graph attention networks (GAT), and the inductive GraphSAGE model. Finally, we introduce the most representative heterogeneous graph neural networks (HAN) for complex graph analysis.
- **Part II**: This part describes the advances of GNNs methods from different aspects. Topics covered in Chaps. 3–7 include Homogeneous Graph Neural Networks, Heterogeneous Graph Neural Networks, Dynamic Graph Neural Networks, Hyperbolic Graph Neural Networks, and Distilling Graph Neural Networks. Each of these topics is covered by a chapter with fundamental concepts and technical details on representative algorithms.

  In Chap. 3, we review some recent works on the homogeneous graph neural networks designing hot topics, covering the work analyzing the feature and topology relations, the work proposing the theoretical framework of GNNs, the work discussing the high-frequency and low-frequency information of GNNs, and the work designing graph structure learning for GNNs. In Chap. 4, we introduce recent heterogeneous graph neural networks considering the propagation depth, distance modeling, adversarial disentangler, and the self-training measure. In Chap. 5, we focus on methods for dynamic graph analysis, providing technical details on dynamic heterogeneous graph neural networks designing. Then, several representative works of hyperbolic graph neural networks are covered in Chap. 6. And we discuss the advanced knowledge distillation methods for graph neural networks in Chap. 7.
- **Part III**: The future direction part describes the advances of methods and applications that tend to be important and promising for future research of graph neural networks. We discuss the advanced topics in GNNs such as robustness, expressiveness, fairness, and so on. Correspondingly, new emerging methods and application domains are also discussed in this part.

Fundamental Graph Neural Networks

2

Mengmei Zhang and Meiqi Zhu

2.1   Introduction

Deep learning has revolutionized many machine learning tasks in recent years, among which Convolutional Neural Networks (CNNs) can extract multi-scale localized spatial features and have high expressiveness, starting the new era of deep learning. Inspired by CNNs, the Graph Convolutional Network (GCN) [102] is proposed to generalize the operation of convolution from grid data to graph data and leads to breakthroughs of graph domain. As the fundamental model for the combination of graph and deep learning, GCN drives the methods applying various neural networks to different graph data, e.g., recurrent graph neural networks [34] (learning node representations with recurrent neural architectures) and graph autoencoders [171] (encoding nodes into a latent vector space via autoencoder architectures for reconstructing graph data). Generally, all these methods are known as Graph Neural Networks (GNNs).

As the basic GNN, GCN bridges the gap between spectral-based (from the perspective of graph signal processing) and spatial-based (from the perspective of information propagation) convolutional GNNs, and points that the essence of GCN is iteratively aggregating the information of neighbors. It inspires a mass of methods redesigning the aggregation process to enhance the adaptability for graph data. In this chapter, we will introduce the fundamental GCN model [102] and its three representative variants: (1) First, we will introduce an inductive graph convolution network framework (GraphSAGE) [74], which learns a function that generates embeddings by sampling and aggregating features from a node's local neighborhood for previously unseen data. (2) Then we will describe the graph atten-

M. Zhang (✉) · M. Zhu
Beijing University of Posts and Telecommunications, Beijing, China

© The Author(s), under exclusive license to Springer Nature Switzerland AG 2023   13
C. Shi et al., *Advances in Graph Neural Networks*, Synthesis Lectures
on Data Mining and Knowledge Discovery,
https://doi.org/10.1007/978-3-031-16174-2_2

tion network (GAT) [189], which extends GCN by introducing attention mechanism for aggregating neighbors, and thus can specify different weights to different nodes in a neighborhood. (3) Finally, in this section, we will introduce the heterogeneous graph attention network (HAN) [200], which extends GCN for heterogeneous graph by leveraging node-level attention and semantic-level attention to learn the importance of nodes and meta-paths. Thus HAN can capture complex structures and rich semantics behind heterogeneous data.

2.2   Graph Convolutional Network

In this section, we introduce the **G**raph **C**onvolutional **N**etwork (GCN) [102], which is the most typical type of graph neural network architectures. It can leverage the graph structure and aggregate node information from the neighborhoods in a convolutional fashion. Graph convolutional network has a great expressive power to learn the graph representations and has achieved a superior performance in a wide range of tasks and applications. Here we introduce the theoretical motivation of GCN, as well as its model design.

2.2.1   Overview

Deep learning models have demonstrated their power in many applications. Especially, Convolution Neural Networks (CNNs) achieve a promising performance in many computer vision applications [133]. One key reason of such successes is that the convolutional layers in CNNs can hierarchically extract high-level features of the images to achieve a great expressive capability by learning a set of fixed-size trainable localized filters.

However, the non-Euclidean characteristic of graphs (e.g., the irregular structure) makes the convolutions and filtering on graphs not as well defined as on images. In the past decades, researchers have been working on how to conduct convolutional operations on graphs. One main research direction is to define graph convolutions from the spectral perspective. Specifically, spectral graph convolutions are defined in the spectral domain based on graph Fourier transform and thus can be computed by taking the inverse Fourier transform of the multiplication between two Fourier transformed graph signals.

Here, the spectral convolutions on graphs are defined as the multiplication of a signal $x \in \mathbb{R}^N$ (a scalar for each node) with a filter $g_\theta = \text{diag}(\theta)$ parameterized by $\theta \in \mathbb{R}^N$ in the Fourier domain, i.e.,

$$g_\theta \star x = U g_\theta U^T x, \tag{2.1}$$

where U is the matrix of eigenvectors of the normalized graph Laplacian $L = I_N - D^{-\frac{1}{2}} A D^{-\frac{1}{2}} = U \Lambda U$, with a diagonal matrix of its eigenvalues Λ and $U^T x$ being the graph Fourier transform of x. We usually design g_θ as a function of the eigenvalues of L, i.e., $g_\theta(\Lambda)$. Evaluating Eq. (2.1) is computationally expensive, as multiplication with the eigenvector matrix U is $O(N^2)$. Furthermore, computing the eigendecomposition of L in the first place might be prohibitively expensive for large graphs. To circumvent this problem, it is suggested in Hammond et al. (2011) that $g_\theta(\Lambda)$ can be well approximated by a truncated expansion in terms of Chebyshev polynomials $T_k(x)$ up to K^{th}-order

$$g_{\theta'}(\Lambda) \approx \sum_{k=0}^{K} \theta'_k T_k(\tilde{\Lambda}), \tag{2.2}$$

with a rescaled $\tilde{\Lambda} = \frac{2}{\lambda_{max}}\Lambda - I_N$. λ_{max} denotes the largest eigenvalue of L. $\theta' \in \mathbb{R}^K$ is now a vector of Chebyshev coefficients. The Chebyshev polynomials are recursively defined as $T_k(x) = 2xT_{k-1}(x) - T_{k-2}(x)$, with $T_0(x) = 1$ and $T_1(x) = x$. The reader is referred to [75] for an in-depth discussion of this approximation.

Going back to our definition of a convolution of a signal x with a filter $g_{\theta'}$, we now have

$$g_{\theta'} \star x \approx \sum_{k=0}^{K} \theta'_k T_k(\tilde{L})x, \tag{2.3}$$

with $\tilde{L} = \frac{2}{\lambda_{max}}L - I_N$. Note that this expression is now K-localized since it is a Kth-order polynomial in the Laplacian, i.e., it depends only on nodes that are at maximum K steps away from the central node (K^{th}-order neighborhood). The complexity of evaluating Eq. (2.3) is $O(|\mathcal{E}|)$, i.e., linear in the number of edges. Reference [36] uses this K-localized convolution to define a convolutional neural network on graphs.

2.2.2   The GCN Method

In GCN model, the authors further simplify the spectral convolutional neural networks on graphs, by truncating the Chebyshev polynomial of convolution operation to first order (i.e., $K = 1$ in Eq. (2.3)) and approximating $\lambda_{max} \approx 2$. These approximations lead to the simplified convolution layer as

$$g_{\theta'} \star x \approx \theta'_0 x + \theta'_1 (L - I_N)x = \theta'_0 x - \theta'_1 D^{-\frac{1}{2}} A D^{-\frac{1}{2}} x, \tag{2.4}$$

with two free parameters θ'_0 and θ'_1. The filter parameters can be shared over the whole graph. This form of filters can be applied successively, then can effectively convolve the Kth-order neighborhood of a node, where K is the number of successive filtering operations or convolutional layers in the neural network model.

In practice, it can be beneficial to restrain the number of parameters to avoid over-fitting and to minimize the number of matrix multiplications per layer. GCN further assumes $\theta = \theta'_0 = -\theta'_1$, leading to the following expression:

$$g_\theta \star x \approx \theta (I_N + D^{-\frac{1}{2}} A D^{-\frac{1}{2}}) x. \tag{2.5}$$

Note that $I_N + D^{-\frac{1}{2}} A D^{-\frac{1}{2}}$ has eigenvalues in the range $[0, 2]$. Repeated application of this operator can therefore lead to numerical instabilities and exploding/vanishing gradients when used in a deep neural network model. To alleviate this problem, we introduce the following renormalization trick: $I_N + D^{-\frac{1}{2}} A D^{-\frac{1}{2}} \to \tilde{D}^{-\frac{1}{2}} \tilde{A} \tilde{D}^{-\frac{1}{2}}$, with $\tilde{A} = A + I_N$ and $\tilde{D}_{ii} = \sum_j \tilde{A}_{ij}$.

We can generalize this definition to a signal $X \in \mathbb{R}^{N \times C}$ with C input channels (i.e., a C-dimensional feature vector for every node) and F filters or feature maps as follows:

$$Z = \tilde{D}^{\frac{1}{2}} \tilde{A} \tilde{D}^{-\frac{1}{2}} X \Theta, \tag{2.6}$$

where $\Theta \in \mathbb{R}^{C \times F}$ is a matrix of filter parameters and $Z \in \mathbb{R}^{N \times F}$ is the convolved signal matrix. This filtering operation has complexity $O(|\mathcal{E}| FC)$, as $\tilde{A} X$ can be efficiently implemented as a product of a sparse matrix with a dense matrix.

Having introduced a simple, yet flexible model $f(X, A)$ for efficient information propagation on graphs, in the following, we consider a two-layer GCN for semi-supervised node classification on a graph with a symmetric adjacency matrix A (binary or weighted) as depicted in Fig. 2.1. We first calculate $\hat{A} = \tilde{D}^{-\frac{1}{2}} \tilde{A} \tilde{D}^{-\frac{1}{2}}$ in a pre-processing step. Our forward model then takes the simple form:

$$Z = f(X, A) = \text{softmax}(\hat{A} \text{ReLU}(\hat{A} X W^{(0)}) W^{(1)}). \tag{2.7}$$

Here, $W^{(0)} \in \mathbb{R}^{C \times H}$ is an input-to-hidden weight matrix for a hidden layer with H feature maps. $W^{(1)} \in \mathbb{R}^{H \times F}$ is a hidden-to-output weight matrix. For semi-supervised multi-class classification, we then evaluate the cross-entropy error over all labeled examples:

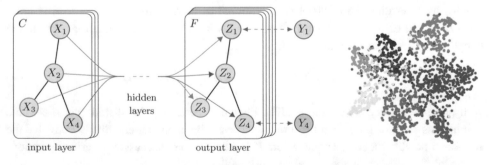

(a) **Graph Convolutional Network**   (b) **Hidden layer activations**

Fig. 2.1 Left: Schematic depiction of multi-layer GCN for semi-supervised learning with C input channels and F feature maps in the output layer. Right: t-SNE visualization of hidden layer activation of a two-layer GCN trained on the Cora dataset using 5% of labels. Colors denote document class

$$\mathcal{L} = - \sum_{l \in \mathbf{Y}_L} \sum_{f=1}^{F} \mathbf{Y}_{lf} \ln \mathbf{Z}_{lf}, \qquad (2.8)$$

where $\mathcal{Y}_L$ is the set of labels and $\mathbf{Y}$ is the label matrix for all nodes. The neural network weights $\mathbf{W}^{(0)}$ and $\mathbf{W}^{(1)}$ are trained using gradient descent. In this work, we perform batch gradient descent using the full dataset for every training iteration, which is a viable option as long as datasets fit in memory. Using a sparse representation for $\mathbf{A}$, memory requirement is $O(|\mathcal{E}|)$, i.e., linear in the number of edges. Stochasticity in the training process is introduced via dropout [174].

2.3   Inductive Graph Convolution Network

In this section, we introduce an aggregation-based inductive graph convolution network, **Graph SA**mple and aggre**GatE** (GraphSAGE) [74]. It is a general inductive framework that learns to generate embeddings by sampling and aggregating features from a node's local neighborhood for previously unseen data. Here we first introduce the background of GraphSAGE, then provide its model framework.

2.3.1   Overview

GCN is designed to learn a good representation of each node from a single fixed graph. However, the training process could be costly in terms of memory for large-scale graphs. Moreover, the main issue of GNNs is that they lack the ability of generalization for unseen data. And it has to re-train the model for the new node in order to represent this node (transductive). Thus, such GNNs are not suitable for dynamic graphs where the nodes in the graphs are ever-changing. Actually, many real-world applications require embeddings to be quickly generated for unseen nodes, or entirely new (sub)graphs. This inductive capability is essential for high-throughput, production machine learning systems, which operate on evolving graphs and constantly encounter unseen nodes (e.g., posts on Reddit, users and videos on YouTube). An inductive approach to generating node embeddings also facilitates generalization across graphs with the same form of features: for example, one could train an embedding generator on protein-protein interaction graphs derived from a model organism, and then easily produce node embeddings for data collected on new organisms using the trained model.

The inductive node embedding problem is especially difficult, compared to the transductive setting, because generalizing to unseen nodes requires aligning newly observed subgraphs to the node embeddings that the algorithm has already optimized on. An inductive framework must learn to recognize structural properties of a node's neighborhood that reveal both the node's local role in the graph and its global position.

2.3.2   The GraphSAGE Method

GraphSAGE extends GCN to the task of inductive unsupervised learning for unseen data, where each node is represented by the aggregation of its neighborhood. Thus, even if a new node unseen during training time appears in the graph, it can still be properly represented by its neighboring nodes. Specifically, on the one hand, instead of full batch training like GCN, GraphSAGE samples local neighbors for each node and then learn how to aggregate feature information from these sampled neighbors in a mini batch way, which is inductive and practical for large-scale graphs. On the other hand, GraphSAGE extends the aggregator of GCN, and proposes a series of alternative operator.

Neighborhood Sampler
We first describe the mechanism of sampling neighbor in GraphSAGE. The input of existing GCN is a fixed size whole graph, and thus the parameters of GCN have to be updated with the gradients of all training samples in each epoch, which is known as full batch learning. Since the whole graph is often large scale in reality, to train in mini batch way, GraphSAGE samples a fixed-size set of neighbors for each node in mini batch, and computes the gradients to update parameters.

GraphSAGE uniformly samples a fixed-size set of neighbors $\mathcal{N}(v)$ at each iteration k, in order to keep the computational footprint of each batch fixed. Without this sampling, the memory and expected runtime of a single batch are unpredictable and in the worst case $O(|\mathcal{V}|)$. As shown in Fig. 2.2, for each node at iteration/depth k, the number of sampled direct neighbors is limited to S_k (a hyper-parameter). Thus the per-batch space and time complexity for GraphSAGE are fixed at $O(\prod_{k=1}^{K} S_k)$, where $S_k, k \in 1, \ldots K$ and K are user-specified constants.

Neighborhood Aggregator
GraphSAGE studies the basic properties of an aggregator: unlike machine learning over sentences and images, a node's neighbors have no natural ordering; thus, the aggregator functions must operate over an unordered set of vectors. Ideally, an aggregator function would be symmetric (i.e., invariant to permutations of its inputs) while still being trainable and maintaining high representational capacity. The symmetry property of the aggregation function ensures that our neural network model can be trained and applied to arbitrarily ordered node neighborhood feature sets.

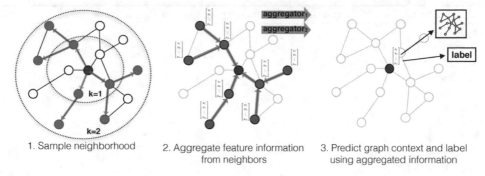

1. Sample neighborhood   2. Aggregate feature information   3. Predict graph context and label
                                from neighbors   using aggregated information

Fig. 2.2 Visual illustration of the GraphSAGE sample and aggregate approach

Then GraphSAGE proposes a series of candidate aggregator functions satisfying the above properties:

Mean aggregator. It takes the element-wise mean of the vectors in $\{\mathbf{h}_u^{k-1}, \forall u \in \mathcal{N}(v)\}$. The mean aggregator is nearly equivalent to the convolutional propagation rule used in the transductive GCN framework. Here we provide the mean aggregator with parameters $\mathbf{W}$ as follows:

$$\mathbf{h}_v^k \leftarrow \sigma(\mathbf{W} \cdot \text{MEAN}(\{\mathbf{h}_v^{k-1}\} \bigcup \{\mathbf{h}_u^{k-1}, \forall u \in \mathcal{N}(v)\})). \tag{2.9}$$

LSTM aggregator. GraphSAGE also examines a more complex aggregator based on an LSTM architecture [80]. Compared to the mean aggregator, LSTMs have the advantage of larger expressive capability. However, it is important to note that LSTMs are not inherently symmetric (i.e., they are not permutation invariant), since they process their inputs in a sequential manner. GraphSAGE adapts LSTMs to operate on an unordered set by simply applying the LSTMs to a random permutation of the node's neighbors.

Pooling aggregator. In pooling approach, the vector of each neighbor is independently fed through a fully connected neural network. Following this transformation, an element-wise max-pooling operation is applied to aggregate information across the neighbor set:

$$AGGREGATE_k^{pool} = max(\{\sigma(\mathbf{W}_{pool}\mathbf{h}_u^k + \mathbf{b}), \forall u \in \mathcal{N}(v)\}). \tag{2.10}$$

GraphSAGE Algorithm

In Algorithm 2.1, we describe the forward propagation process of mini batch embedding generation.

Algorithm 2.1 GraphSAGE mini batch forward propagation algorithm

---
1: $\mathcal{B}^K \leftarrow \mathcal{B}$;
2: **for** $k = K, \ldots, 1$ **do**
3:     $\mathcal{B}^{k-1} \leftarrow \mathcal{B}^k$;
4:     **for** $v \in \mathcal{B}^k$ **do**
5:         $\mathcal{B}^{k-1} \leftarrow \mathcal{B}^k \bigcup \mathcal{N}_k(v)$;
6:     **end for**
7: **end for**
8: $\mathbf{h}_v^0 \leftarrow \mathbf{x}_v, \forall v \in \mathcal{B}^0$;
9: **for** $k = 1, \ldots, K$ **do**
10:     **for** $v \in \mathcal{B}^k$ **do**
11:         $\mathbf{h}_{\mathcal{N}(v)}^k \leftarrow \text{AGGREGATE}_k(\{\mathbf{h}_u^{k-1}, \forall u \in \mathcal{N}_k(v)\})$;
12:         $\mathbf{h}_v^k \leftarrow \sigma(\mathbf{W}^k \cdot \text{CONCAT}(\mathbf{h}_v^{k-1}), \mathbf{h}_{\mathcal{N}(v)}^k)$;
13:         $\mathbf{h}_v^k \leftarrow \mathbf{h}_v^k / \|\mathbf{h}_v^k\|_2$;
14:     **end for**
15: **end for**
16: $\mathbf{z}_v \leftarrow \mathbf{h}_v^K, \forall v \in \mathcal{B}$;

Given graph $\mathcal{G} = (\mathcal{V}, \mathcal{E})$ and features for all nodes $\mathbf{x}_v, \forall v \in \mathcal{V}$, the main idea is to sample all the nodes needed for the computation first. Lines 2–7 of Algorithm 2.1 correspond to the sampling stage. Each set $\mathcal{B}^K$ contains the nodes that are needed to compute the representations of nodes $v \in \mathcal{B}^{k+1}$, i.e., the nodes in the $(k+1)$-st iteration/layer. Lines 9–15 correspond to the aggregation stage. Each step in the outer loop of aggregation stage proceeds as follows, where k denotes the current step in the outer loop (or the depth of the search) and $\mathbf{h}^k$ denotes a node's representation at this step: (1) Each node $v \in \mathcal{V}$ aggregates the representations of the nodes in its immediate neighborhood, $\{\mathbf{h}_u^{k-1}, \forall u \in \mathcal{N}(v)\}$, into a single vector $\mathbf{h}_{\mathcal{N}(v)}^{k-1}$. The $k = 0$ representations are defined as the input node features. (2) Then GraphSAGE concatenates the node's current representation, $\mathbf{h}_v^{k-1}$, with the aggregated neighborhood vector, $\mathbf{h}_{\mathcal{N}(v)}^{k-1}$, and this concatenated vector is fed through a fully connected layer with non-linear activation function σ, yielding $\mathbf{h}_v^k$. For notational convenience, we denote the final representations output at depth K as $\mathbf{z}_v \leftarrow \mathbf{h}_v^K$.

2.4   Graph Attention Network

In this section, we introduce the **Graph AT**tention network (GAT) [189], which extends GCN by introducing attention mechanism for aggregating neighbors. Thus GAT can specify different weights to different nodes in a neighborhood and have better expressive power.

2.4.1   Overview

Attention mechanism in deep neural networks is a technique that mimics cognitive attention. It is inspired by the human's cognitive process of focusing selectively on certain aspect of information, while ignoring other perceptible information. The effect enhances the important parts of the input data and fades out the rest—the thought being that the network should devote more computing power to that small but important part of the data. Which part of the data is more important than others depends on the context and is learned through training data by gradient descent.

For example in translation task, the goal is to translate the input sentence "How was your day" to the French version—"Comment se passe ta journée". For each word in the output sentence, the attention mechanism in neural networks will assign more attention to important and relevant words from the input sentence and assign higher weights to these words, enhancing the accuracy of the output prediction.

Similarly, in the real world, graphs can be both large—with many complex patterns—and noisy which can pose a problem for effective graph mining. An effective way to deal with this issue is to incorporate "attention" into graph mining solutions. An attention mechanism allows a method to focus on task-relevant parts of the graph, helping it to make better decisions.

2.4.2   The GAT Method

We will start by describing a single graph attentional layer, as the sole layer utilized throughout all of the GAT architectures. The input to our layer is a set of node features, $\{\mathbf{h}_1, \ldots, \mathbf{h}_N\}$, $\mathbf{h}_i \in \mathbb{R}^F$, where N is the number of nodes, and F is the number of features in each node. The layer produces a new set of node features (of potentially different cardinality F'), $\{\mathbf{h}'_1, \ldots, \mathbf{h}'_N\}$, $\mathbf{h}'_i \in \mathbb{R}^{F'}$, as its output.

In order to obtain sufficient expressive power to transform the input features into higher level features, at least one learnable linear transformation is required. To that end, as an initial step, a shared linear transformation, parametrized by a weight matrix, $\mathbf{W} \in \mathbb{R}^{F' \times F}$, is applied to every node. We then perform a shared self-attention (i.e., attn : $\mathbb{R}^{F'} \times \mathbb{R}^{F'} \to \mathbb{R}$) on the nodes to compute attention coefficients:

$$e_{ij} = \text{attn}(\mathbf{W}\mathbf{h}_i, \mathbf{W}\mathbf{h}_j), \tag{2.11}$$

which indicates the importance of node j's features to node i. In its most general formulation, the model allows every node to attend on every other node, dropping all structural information. We inject the graph structure into the mechanism by performing masked attention, and we only compute e_{ij} for nodes $j \in \mathcal{N}_i$, where $\mathcal{N}_i$ is the neighbors of node i in the graph. In all our experiments, these will be exactly the first-order neighbors of i (including

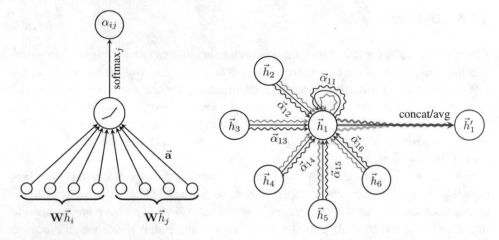

Fig. 2.3 Left: The attention mechanism attn $(\mathbf{Wh}_i, \mathbf{Wh}_j)$ employed by GAT model. Right: An illustration of multi-head attention (with $K = 3$ heads) by node 1 on its neighborhood. Different arrow styles and colors denote independent attention computations

i). To make coefficients easily comparable across different nodes, we normalize them across all choices of j using the softmax function:

$$\alpha_{ij} = \text{softmax}_j(e_{ij}) = \frac{\exp(e_{ij})}{\sum_{k \in \mathcal{N}_i} \exp(e_{ik})}. \tag{2.12}$$

In our experiments, the attention mechanism attn is a single-layer feedforward neural network, parametrized by a weight vector $\mathbf{a} \in \mathbb{R}^{2F'}$, and applying the LeakyReLU nonlinearity (with negative input slope $\alpha = 0.2$). Fully expanded out, the coefficients computed by the attention mechanism (illustrated by Fig. 2.3 (left)) can then be expressed as

$$\alpha_{ij} = \frac{\exp(\text{LeakyReLU}(\mathbf{a}^T[\mathbf{Wh}_i \| \mathbf{Wh}_j]))}{\sum_{k \in \mathcal{N}_i} \exp(\text{LeakyReLU}(\mathbf{a}^T[\mathbf{Wh}_i \| \mathbf{Wh}_k]))}, \tag{2.13}$$

where $\|$ is the concatenation operation.

Once obtained, the normalized attention coefficients are used to compute a linear combination of the features corresponding to them, to serve as the final output features for every node (after potentially applying a non-linearity, σ):

$$\mathbf{h}_i = \sigma\left(\sum_{j \in \mathcal{N}_i} \alpha_{ij} \mathbf{Wh}_j\right). \tag{2.14}$$

To stabilize the learning process of self-attention, we have found extending our mechanism to employ multi-head attention to be beneficial, similar to [187]. Specifically, K

independent attention mechanisms execute the transformation of Eq. (2.14), and then their features are concatenated, resulting in the following output feature representation:

$$\mathbf{h}_i = \|_{k=1}^{K} \sigma \left(\sum_{j \in \mathcal{N}_i} \alpha_{ij}^k \mathbf{W}^k \mathbf{h}_j \right), \tag{2.15}$$

where α_{ij}^k are normalized attention coefficients computed by the k-th attention mechanism (attn^k), and $\mathbf{W}^k$ is the corresponding input linear transformation's weight matrix. Note that, in this setting, the final returned output, $\mathbf{h}'$, will consist of KF' features (rather than F') for each node.

Specially, if we perform multi-head attention on the final (prediction) layer of the network, concatenation is no longer sensible—instead, we employ averaging, and delay applying the final non-linearity (usually a softmax or logistic sigmoid for classification problems) until then:

$$\mathbf{h}_i = \sigma \left(\frac{1}{K} \sum_{k=1}^{K} \sum_{j \in \mathcal{N}_i} \alpha_{ij}^k \mathbf{W}^k \mathbf{h}_j \right). \tag{2.16}$$

The aggregation process of a multi-head graph attentional layer is illustrated in Fig. 2.3 (right).

Overall, as opposed to GCN, we allow for (implicitly) assigning different importance to nodes of a same neighborhood, enabling a leap in model capacity. Furthermore, analyzing the learned attentional weights may lead to benefits in interpretability. The attention mechanism is applied in a shared manner to all edges in the graph, and therefore it does not depend on upfront access to the global graph structure or (features of) all of its nodes (a limitation of many prior techniques).

2.5 Heterogeneous Graph Attention Network

In this section, we introduce the **H**eterogeneous graph **A**ttention **N**etwork (HAN) [200], which extends GCN for heterogeneous graph by leveraging node-level attention and semantic-level attention to learn the importance of nodes and meta-paths. Thus HAN can capture complex structures and rich semantics behind heterogeneous.

2.5.1 Overview

Despite the success of attention mechanism in deep learning, it has not been considered in the graph neural network framework for heterogeneous graph, which contains multi-types of nodes and edges. It is important since the heterogeneous graph contains more comprehensive information and rich semantics, and it has been widely used in many data mining tasks. Due to the complexity of heterogeneous graph, traditional graph neural networks cannot be directly applied to heterogeneous graph.

Specifically, the graph neural network architecture for heterogeneous graph should address the following requirements:

Heterogeneity of graph. The heterogeneity is an intrinsic property of heterogeneous graph, i.e., various types of nodes and edges. For example, different types of nodes have different traits and their features may fall in different feature spaces.

Semantic-level attention. Different meaningful and complex semantic information are involved in heterogeneous graph, which are usually reflected by meta-paths [179]. Different meta-paths in heterogeneous graph may extract diverse semantic information. How to select the most meaningful meta-paths and fuse the semantic information for the specific task is an open problem [164]. Semantic-level attention aims to learn the importance of each meta-path and assign proper weights to them.

Node-level attention. In a heterogeneous graph, nodes can be connected via various types of relations, e.g., meta-path. Given a meta-path, each node has lots of meta-path-based neighbors. How to distinguish the subtle difference of their neighbors and select some informative neighbors is required. For each node, node-level attention aims to learn the importance of meta-path-based neighbors and assign different attention values to them.

2.5.2   The HAN Method

In this section, we describe HAN following its hierarchical attention structure: node-level attention and semantic-level attention. Figure 2.4 presents the whole framework of HAN. First, HAN utilizes a node-level attention to learn the weight of meta-path-based neighbors and aggregates them to get the semantic-specific node embedding. After that, HAN can tell the difference of meta-paths via semantic-level attention and get the optimal weighted combination of the semantic-specific node embedding for the specific task.

Node-level Attention
Here we first describe node-level attention, which can learn the importance of meta-path-based neighbors for each node in a heterogeneous graph and aggregate the representation of these meaningful neighbors to form a node embedding.

Due to the heterogeneity of nodes, different types of nodes have different feature spaces. HAN projects each type of nodes (e.g., node with type ϕ_i) to same space with type-specific transformation matrix $\mathbf{M}_{\phi_i}$ as follows:

$$\mathbf{h}'_i = \mathbf{M}_{\phi_i} \cdot \mathbf{h}_i, \tag{2.17}$$

where $\mathbf{h}_i$ and $\mathbf{h}'_i$ are the original and projected feature of node i, respectively.

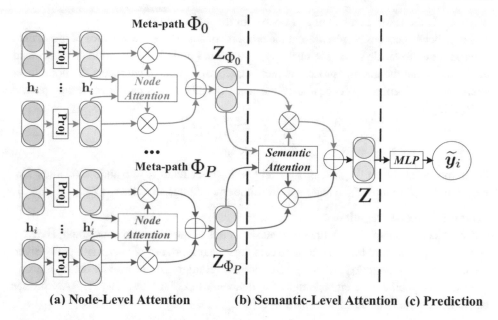

(a) Node-Level Attention   (b) Semantic-Level Attention (c) Prediction

Fig. 2.4 The overall framework of the proposed HAN

After that, given a node pair (i, j) which is connected via meta-path Φ, the node-level attention mechanism of HAN will compute the importance e_{ij}^{Φ} which means how important node j will be for node i as follows:

$$e_{ij}^{\Phi} = att_{node}(\mathbf{h}'_i, \mathbf{h}'_j; \Phi). \tag{2.18}$$

Here att_{node} denotes the DNN which performs the node-level attention. Then for each node i, HAN normalizes e_{ij}^{Φ} for all meta-path-based neighbors $j \in \mathcal{N}_i^{\Phi}$, yielding the weight coefficient α_{ij}^{Φ}:

$$\alpha_{ij}^{\Phi} = softmax_j(e_{ij}^{\Phi}) = \frac{\exp\left(\sigma(\mathbf{a}_{\Phi}^{\mathrm{T}} \cdot [\mathbf{h}'_i \| \mathbf{h}'_j])\right)}{\sum_{k \in \mathcal{N}_i^{\Phi}} \exp\left(\sigma(\mathbf{a}_{\Phi}^{\mathrm{T}} \cdot [\mathbf{h}'_i \| \mathbf{h}'_k])\right)}, \tag{2.19}$$

where $\|$ denotes the concatenate operation and $\mathbf{a}_{\Phi}$ is the node-level attention vector for meta-path Φ. As we can see, the weight coefficient of (i, j) depends on their features. Then, the meta-path-based embedding of node i can be aggregated by the neighbor's projected features with the corresponding coefficients as follows:

$$\mathbf{z}_i^{\Phi} = \sigma\left(\sum_{j \in \mathcal{N}_i^{\Phi}} \alpha_{ij}^{\Phi} \cdot \mathbf{h}'_j\right), \tag{2.20}$$

where $\mathbf{z}_i^\Phi$ is the learned embedding of node i for the meta-path Φ.

Since heterogeneous graphs present the property of scale free, the variance of graph data is quite high. To tackle the above challenge, we extend node-level attention to multi-head attention so that the training process is more stable. Specifically, we repeat the node-level attention for K times and concatenate the learned embeddings as the semantic-specific embedding:

$$\mathbf{z}_i^\Phi = \overset{K}{\underset{k=1}{\|}} \sigma \left(\sum_{j \in \mathcal{N}_i^\Phi} \alpha_{ij}^\Phi \cdot \mathbf{h}_j' \right). \tag{2.21}$$

Given the meta-path set $\{\Phi_1, \dots, \Phi_P\}$, after feeding node features into node-level attention, we can obtain P groups of semantic-specific node embeddings, denoted as $\{\mathbf{Z}_{\Phi_1}, \dots, \mathbf{Z}_{\Phi_P}\}$.

Semantic-level Aggregation
Since different meta-paths capture different semantics of the heterogeneous graph, HGNNs usually adopt semantic-level attention to calculate the importance of each meta-path. Given the meta-path set $\{\Phi_0, \Phi_1, \dots, \Phi_P\}$, after node-level aggregation, we can obtain a group of semantic-specific node embeddings of v, denoted as $\{\mathbf{z}_v^{\Phi_0}, \mathbf{z}_v^{\Phi_1}, \dots, \mathbf{z}_v^{\Phi_P}\}$. HAN further calculates the importance of meta-path $\Phi \in \{\Phi_1, \dots, \Phi_P\}$ by

$$w^\Phi = \frac{1}{|\mathcal{V}|} \sum_{v \in \mathcal{V}} \mathbf{q}^T \cdot tanh(\mathbf{W} \cdot \mathbf{z}_v^\Phi + \mathbf{b}), \tag{2.22}$$

where $\mathbf{W}$ and $\mathbf{b}$ denote the weight matrix and bias of the MLP, respectively. $\mathbf{q}$ is the semantic-level attention vector. Then HAN uses the softmax function to normalize the importance w^Φ to yield the attention value β^Φ for Φ. Hence, the final embedding $\mathbf{z}_v$ of v can be obtained by semantic-level aggregation:

$$\mathbf{z}_v = \sum_{\Phi \in \{\Phi_1, \dots, \Phi_P\}} \beta^\Phi \cdot \mathbf{z}_v^\Phi. \tag{2.23}$$

Finally, the overall proposed model can be optimized by minimizing following loss:

$$\mathcal{L} = - \sum_{v \in \mathcal{V}_L} \ln(\mathbf{W}_{clf} \cdot \mathbf{z}_{v,c_v}), \tag{2.24}$$

where $\mathbf{W}_{clf}$ is the parameter of the classifier, c_v is the class of training node $v \in \mathcal{V}_L$. With the guide of labeled data, we can optimize the proposed model via backpropagation and learn the embeddings of nodes.

Homogeneous Graph Neural Networks

3

Deyu Bo

3.1 Introduction

Network is a ubiquitous structure for real-world data, such as social networks, citation networks, and financial networks. Recently, Graph Neural Networks (GNNs) have gained great popularity in tackling analytic tasks on graph-structured data. The well-designed message-passing (or propagation) mechanism is the most fundamental part of GNNs. For example, Graph Convolutional Networks (GCNs) [102] use degrees to normalize the information of neighbors, Graph Attention Networks (GATs) [189] apply the attention mechanism to GNNs to find important neighbors, and GraphSAGE [74] employs the mean and max-pooling strategies. Despite their success, there are still some problems in existing message-passing functions, which may lead to sub-optimal performance in some applications.

In Sect. 3.2, we first introduce an observation that GCNs is not expert in fusing node features and topological structures. Therefore, a new GNN, namely, Adaptive Multi-channel Graph Convolutional Networks (AM-GCN), is proposed to adaptively aggregate the feature and structure information in the message-passing process. Section 3.3 introduces Frequency Adaptation Graph Convolutional Networks (FAGCN), which can adaptively aggregate the low-frequency and high-frequency information. FAGCN designs a generalized attention mechanism that can help existing message-passing methods get rid of the low-pass filtering. Section 3.4 recommends Graph Estimation Neural Networks (GEN), which can learn a better message-passing structure, i.e., graph topology, for GNNs. GEN is more robust than GCNs due to its powerful ability in denoising and community detecting. Besides, in Sect. 3.5, we introduce a unified framework of existing GNNs, which summarizes different message-

D. Bo (✉)
Beijing University of Posts and Telecommunications, Beijing, China

© The Author(s), under exclusive license to Springer Nature Switzerland AG 2023
C. Shi et al., *Advances in Graph Neural Networks*, Synthesis Lectures
on Data Mining and Knowledge Discovery,
https://doi.org/10.1007/978-3-031-16174-2_3

passing functions into a closed-form objective. This discovery can help the researchers understand the principles behind the message-passing mechanism. Finally, Sect. 3.6 gives a comprehensive conclusion of this chapter.

3.2   Adaptive Multi-channel Graph Convolutional Networks

3.2.1   Overview

The enormous success of GCNs is partially due to the reason that GCNs provide a fusion strategy on topological structures and node features to learn node embedding, and the fusion process is supervised by an end-to-end learning framework. Some recent studies, however, disclose certain weakness of the state-of-the-art GCNs in fusing node features and topological structures [116, 145, 207]. What information do GCNs really learn and fuse from topological structures and node features? This is a fundamental question since GCNs are often used as an end-to-end learning framework. A well-informed answer to this question can help us understand the capability and limitations of GCNs in a principled way. This motivates our study immediately.

As the first contribution of this study, we present experiments assessing the capability of GCNs in fusing topological structures and node features. Surprisingly, our experiments clearly show that the fusion capability of GCNs on network topological structures and node features is clearly distant from optimal or even satisfactory. Even under some simple situations that the correlation between node features/topology with node label is very clear, GCNs still cannot adequately fuse node features and topological structures to extract the most correlated information. The weakness may severely hinder the capability of GCNs in some classification tasks, since GCNs may not be able to adaptively learn some correlation information between topological structures and node features.

To address the weakness of the state-of-the-art GCNs in fusion node features and graph topology, we propose an adaptive multi-channel graph convolutional network for semi-supervised classification (AM-GCN). The central idea is that we learn the node embedding based on node features, topological structures, and their combinations simultaneously. The rationale is that the similarity between features and that inferred by topological structures is complementary to each other and can be fused adaptively to derive deeper correlation information for classification tasks. Experimental results on a series of benchmark datasets clearly show that AM-GCN outperforms the state-of-the-art GCNs and extracts the most correlation information from both node features and topological structures nicely for challenging classification tasks.

3.2.2   Investigation

We use two simple yet intuitive cases to examine whether the state-of-the-art GCNs can adaptively learn from node features and topological structures in graphs and fuse them sufficiently for classification tasks. The main idea is that we will clearly establish the high correlation between node label with network topology and node features, respectively, then we will check the performance of GCN on these two simple cases. A good fusion capability of GCN should adaptively extract the correlated information with the supervision of node label, providing a good result.

Case 1: Random Topology and Correlated Node Features. We generate a random network consisting of 900 nodes, where the probability of building an edge between any two nodes is 0.03. Each node has a feature vector of 50 dimensions. To generate node features, we randomly assign 3 labels to the 900 nodes, and for the nodes with the same label, we use one Gaussian distribution to generate the node features. The Gaussian distributions for the three classes of nodes have the same covariance matrix, but three different centers far away from each other. In this dataset, the node labels are highly correlated with the node features, but not the topological structures. We apply GCN [102] to train this network. For each class, we randomly select 20 nodes for training and another 200 nodes for testing. We carefully tune the hyper-parameters to report the best performance and avoid over-smoothing. Also, we apply MLP [146] to the node features only. The classification accuracies of GCN and MLP are 75.2% and 100%, respectively.

The results meet the expectation. Since the node features are highly correlated with the node labels, MLP shows excellent performance. GCN extracts information from both the node features and the topological structures, but cannot adaptively fuse them to avoid the interference from topological structures. It cannot match the high performance of MLP.

Case 2: Correlated Topology and Random Node Features. We generate another network with 900 nodes. This time, the node features, each of 50 dimensions, are randomly generated. For the topological structure, we employ the Stochastic Blockmodel (SBM) [98] to split nodes into 3 communities (nodes 0–299, 300–599, 600–899, respectively). Within each community, the probability of building an edge is set to 0.03, and the probability of building an edge between nodes in different communities is set to 0.0015. In this dataset, the node labels are determined by the communities, i.e., nodes in the same community have the same label. Again, we apply GCN to this network. We also apply DeepWalk [154] to the topology of the network, that is, the features are ignored by DeepWalk. The classification accuracies of GCN and DeepWalk are 87% and 100%, respectively.

DeepWalk performs well because it models network topological structures thoroughly. GCN extracts information from both the node features and the topological structures, but cannot adaptively fuse them to avoid the interference from node features. It cannot match the high performance of DeepWalk.

Summary. These cases show that the current fusion mechanism of GCN [102] is distant from optimal or even satisfactory. Even though the correlation between node labels with network topology or node features is very high, the current GCN cannot make full use of the supervision by node labels to adaptively extract the most correlated information. However, the situation is more complex in reality, because it is hard to know whether the topology or the node features are more correlated with the final task, which prompts us to rethink the current mechanism of GCN.

3.2.3   The AM-GCN Method

We focus on semi-supervised node classification in an attributed graph $G = (\mathbf{A}, \mathbf{X})$, where $\mathbf{A} \in \mathbb{R}^{n \times n}$ is the symmetric adjacency matrix with n nodes and $\mathbf{X} \in \mathbb{R}^{n \times d}$ is the node feature matrix, and d is the dimension of node features. Specifically, $A_{ij} = 1$ represents there is an edge between nodes i and j, otherwise, $A_{ij} = 0$. We suppose each node belongs to one out of C classes.

3.2.3.1 Overall Framework of AM-GCN

The overall framework of AM-GCN is shown in Fig. 3.1. The key idea of AM-GCN is to permit node features to propagate not only in topology space, but also in feature space, and the most correlated information with node label should be extracted from both of these two

Fig. 3.1 The framework of AM-GCN model. Node feature $\mathbf{X}$ is to construct a feature graph. AM-GCN consists of two specific convolution modules, one common convolution module and the attention mechanism

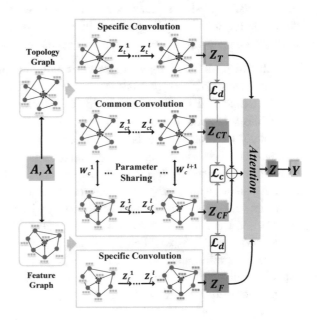

spaces. To this end, we construct a feature graph based on node features $\mathbf{X}$. Then with two specific convolution modules, $\mathbf{X}$ is able to propagate over both of feature graph and topology graph to learn two specific embeddings $\mathbf{Z}_F$ and $\mathbf{Z}_T$, respectively. Further, considering that the information in these two spaces has common characteristics, we design a common convolution module with parameter sharing strategy to learn the common embedding $\mathbf{Z}_{CF}$ and $\mathbf{Z}_{CT}$, also, a consistency constraint $\mathcal{L}_c$ is employed to enhance the "common" property of $\mathbf{Z}_{CF}$ and $\mathbf{Z}_{CT}$. Besides, a disparity constraint $\mathcal{L}_d$ is to ensure the independence between $\mathbf{Z}_F$ and $\mathbf{Z}_{CF}$, as well as $\mathbf{Z}_T$ and $\mathbf{Z}_{CT}$. Considering that node label may be correlated with topology or feature or both, AM-GCN utilizes an attention mechanism to adaptively fuse these embeddings with the learned weights, so as to extract the most correlated information $\mathbf{Z}$ for the final classification task.

3.2.3.2 Specific Convolution Module

Firstly, in order to capture the underlying structure of nodes in feature space, we construct a k-nearest neighbor (kNN) graph $G_f = (\mathbf{A}_f, \mathbf{X})$ based on node feature matrix $\mathbf{X}$, where $\mathbf{A}_f$ is the adjacency matrix of kNN graph. Specifically, we first calculate the similarity matrix $\mathbf{S} \in \mathbb{R}^{n \times n}$ among n nodes. Actually, there are many ways to obtain $\mathbf{S}$, and we list two popular ones here, in which $\mathbf{x}_i$ and $\mathbf{x}_j$ are feature vectors of nodes i and j:

(1) **Cosine Similarity**: It uses the cosine value of the angle between two vectors to measure the similarity:

$$\mathbf{S}_{ij} = \frac{\mathbf{x}_i \cdot \mathbf{x}_j}{|\mathbf{x}_i||\mathbf{x}_j|}. \tag{3.1}$$

(2) **Heat Kernel**: The similarity is calculated by Eq. (3.2) where t is the time parameter in heat conduction equation and we set $t = 2$.

$$\mathbf{S}_{ij} = e^{-\frac{\|\mathbf{x}_i - \mathbf{x}_j\|^2}{t}}. \tag{3.2}$$

Here we uniformly choose the Cosine similarity to obtain the similarity matrix $\mathbf{S}$, and then we choose top k similar node pairs for each node to set edges and finally get the adjacency matrix $\mathbf{A}_f$.

Then with the input graph $(\mathbf{A}_f, \mathbf{X})$ in feature space, the l-th layer output $\mathbf{Z}_f^{(l)}$ can be represented as

$$\mathbf{Z}_f^{(l)} = ReLU(\tilde{\mathbf{D}}_f^{-\frac{1}{2}} \tilde{\mathbf{A}}_f \tilde{\mathbf{D}}_f^{-\frac{1}{2}} \mathbf{Z}_f^{(l-1)} \mathbf{W}_f^{(l)}), \tag{3.3}$$

where $\mathbf{W}_f^{(l)}$ is the weight matrix of the l-th layer in GCN, $ReLU$ is the Relu activation function and the initial $\mathbf{Z}_f^{(0)} = \mathbf{X}$. Specifically, we have $\tilde{\mathbf{A}}_f = \mathbf{A}_f + \mathbf{I}_f$ and $\tilde{\mathbf{D}}_f$ is the diagonal degree matrix of $\tilde{\mathbf{A}}_f$. We denote the last layer output embedding as $\mathbf{Z}_F$. In this way, we can learn the node embedding which captures the specific information $\mathbf{Z}_F$ in feature space.

As for the topology space, we have the original input graph $G_t = (\mathbf{A}_t, \mathbf{X}_t)$ where $\mathbf{A}_t = \mathbf{A}$ and $\mathbf{X}_t = \mathbf{X}$. Then the learned output embedding $\mathbf{Z}_T$ based on topology graph can be calculated in the same way as in feature space. Therefore, the specific information encoded in topology space can be extracted.

3.2.3.3 Common Convolution Module

In reality, the feature and topology spaces are not completely irrelevant. Basically, node classification task may be correlated with the information either in feature space or in topology space or in both of them, which is difficult to know beforehand. Therefore, we not only need to extract the node-specific embedding in these two spaces, but also to extract the common information shared by the two spaces. In this way, it will become more flexible for the task to determine which part of information is the most correlated. To address this, we design a *Common*-GCN with parameter sharing to get the embedding shared in two spaces.

First, we utilize *Common*-GCN to extract the node embedding $\mathbf{Z}_{ct}^{(l)}$ from topology graph $(\mathbf{A}_t, \mathbf{X})$ as follows:

$$\mathbf{Z}_{ct}^{(l)} = ReLU(\tilde{\mathbf{D}}_t^{-\frac{1}{2}}\tilde{\mathbf{A}}_t\tilde{\mathbf{D}}_t^{-\frac{1}{2}}\mathbf{Z}_{ct}^{(l-1)}\mathbf{W}_c^{(l)}), \tag{3.4}$$

where $\mathbf{W}_c^{(l)}$ is the l-th layer weight matrix of *Common*-GCN and $\mathbf{Z}_{ct}^{(l-1)}$ is the node embedding in the $(l-1)$-th layer and $\mathbf{Z}_{ct}^{(0)} = \mathbf{X}$. When utilizing *Common*-GCN to learn the node embedding from feature graph $(\mathbf{A}_f, \mathbf{X})$, in order to extract the shared information, we share the same weight matrix $\mathbf{W}_c^{(l)}$ for every layer of *Common*-GCN:

$$\mathbf{Z}_{cf}^{(l)} = ReLU(\tilde{\mathbf{D}}_f^{-\frac{1}{2}}\tilde{\mathbf{A}}_f\tilde{\mathbf{D}}_f^{-\frac{1}{2}}\mathbf{Z}_{cf}^{(l-1)}\mathbf{W}_c^{(l)}), \tag{3.5}$$

where $\mathbf{Z}_{cf}^{(l)}$ is the l-layer output embedding and $\mathbf{Z}_{cf}^{(0)} = \mathbf{X}$. The shared weight matrix can filter out the shared characteristics from two spaces. According to different input graphs, we can get two output embedding $\mathbf{Z}_{CT}$ and $\mathbf{Z}_{CF}$ and the common embedding $\mathbf{Z}_C$ of the two spaces is

$$\mathbf{Z}_C = (\mathbf{Z}_{CT} + \mathbf{Z}_{CF})/2. \tag{3.6}$$

3.2.3.4 Attention Module

Now we have two specific embeddings $\mathbf{Z}_T$ and $\mathbf{Z}_F$, and one common embedding $\mathbf{Z}_C$. Considering the node label can be correlated with one of them or even their combinations, we use the attention mechanism $att(\mathbf{Z}_T, \mathbf{Z}_C, \mathbf{Z}_F)$ to learn their corresponding importance $(\alpha_t, \alpha_c, \alpha_f)$ as follows:

$$(\alpha_t, \alpha_c, \alpha_f) = att(\mathbf{Z}_T, \mathbf{Z}_C, \mathbf{Z}_F), \tag{3.7}$$

where $\alpha_t, \alpha_c, \alpha_f \in \mathbb{R}^{n \times 1}$ indicate the attention values of n nodes with embeddings $\mathbf{Z}_T, \mathbf{Z}_C, \mathbf{Z}_F$, respectively.

Here we focus on node i, where its embedding in $\mathbf{Z}_T$ is $\mathbf{z}_T^i \in \mathbb{R}^{1 \times h}$ (i.e., the i-th row of $\mathbf{Z}_T$). We firstly transform the embedding through a non-linear transformation, and then use one shared attention vector $\mathbf{q} \in \mathbb{R}^{h' \times 1}$ to get the attention value ω_T^i:

$$\omega_T^i = \mathbf{q}^T \cdot tanh(\mathbf{W}_T \cdot (\mathbf{z}_T^i)^T + \mathbf{b}_T). \tag{3.8}$$

Here $\mathbf{W}_T \in \mathbb{R}^{h' \times h}$ is the weight matrix and $\mathbf{b}_T \in \mathbb{R}^{h' \times 1}$ is the bias vector for embedding matrix $\mathbf{Z}_T$, respectively. Similarly, we can get the attention values ω_C^i and ω_F^i for node i in embedding matrices $\mathbf{Z}_C$ and $\mathbf{Z}_F$, respectively. We then normalize the attention values $\omega_T^i, \omega_C^i, \omega_F^i$ with softmax function to get the final weight:

$$\alpha_T^i = softmax(\omega_T^i) = \frac{exp(\omega_T^i)}{exp(\omega_T^i) + exp(\omega_C^i) + exp(\omega_F^i)}. \tag{3.9}$$

Larger α_T^i implies the corresponding embedding is more important. Similarly, $\alpha_C^i = softmax(\omega_C^i)$ and $\alpha_F^i = softmax(\omega_F^i)$. For all the n nodes, we have the learned weights $\alpha_t = [\alpha_T^i], \alpha_c = [\alpha_C^i], \alpha_f = [\alpha_F^i] \in \mathbb{R}^{n \times 1}$, and denote $\alpha_T = diag(\alpha_t), \alpha_C = diag(\alpha_c)$, and $\alpha_F = diag(\alpha_f)$. Then we combine them to obtain the final embedding $\mathbf{Z}$:

$$\mathbf{Z} = \alpha_T \cdot \mathbf{Z}_T + \alpha_C \cdot \mathbf{Z}_C + \alpha_F \cdot \mathbf{Z}_F. \tag{3.10}$$

3.2.3.5 Objective Function

For the two output embeddings $\mathbf{Z}_{CT}$ and $\mathbf{Z}_{CF}$ of *Common*-GCN, despite the *Common*-GCN has the shared weight matrix, here we design a consistency constraint to further enhance their commonality.

Firstly, we use L_2-normalization to normalize the embedding matrix as $\mathbf{Z}_{CTnor}, \mathbf{Z}_{CFnor}$. Then, the two normalized matrices can be used to capture the similarity of n nodes as $\mathbf{S}_T$ and $\mathbf{S}_F$ as follows:

$$\begin{aligned} \mathbf{S}_T &= \mathbf{Z}_{CTnor} \cdot \mathbf{Z}_{CTnor}^T, \\ \mathbf{S}_F &= \mathbf{Z}_{CFnor} \cdot \mathbf{Z}_{CFnor}^T. \end{aligned} \tag{3.11}$$

The consistency implies that the two similarity matrices should be similar, which gives rise to the following constraint:

$$\mathcal{L}_c = \|\mathbf{S}_T - \mathbf{S}_F\|_F^2. \tag{3.12}$$

Here because embeddings $\mathbf{Z}_T$ and $\mathbf{Z}_{CT}$ are learned from the same graph $G_t = (\mathbf{A}_t, \mathbf{X}_t)$, to ensure they can capture different information, we employ the Hilbert-Schmidt Independence Criterion (HSIC) [172], a simple but effective measure of independence, to enhance the disparity of these two embeddings. Due to its simplicity and neat theoretical properties, HSIC has been applied to several machine learning tasks [68, 144]. Formally, the HSIC constraint of $\mathbf{Z}_T$ and $\mathbf{Z}_{CT}$ is defined as

$$HSIC(\mathbf{Z}_T, \mathbf{Z}_{CT}) = (n-1)^{-2} tr(\mathbf{RK}_T \mathbf{RK}_{CT}), \tag{3.13}$$

where $\mathbf{K}_T$ and $\mathbf{K}_{CT}$ are the Gram matrices with $k_{T,ij} = k_T(\mathbf{z}_T^i, \mathbf{z}_T^j)$ and $k_{CT,ij} = k_{CT}(\mathbf{z}_{CT}^i, \mathbf{z}_{CT}^j)$. And $\mathbf{R} = \mathbf{I} - \frac{1}{n}ee^T$, where $\mathbf{I}$ is an identity matrix and e is an all-one column vector. In this Section, we use the inner product function for $\mathbf{K}_T$ and $\mathbf{K}_{CT}$.

Similarly, considering the embeddings $\mathbf{Z}_F$ and $\mathbf{Z}_{CF}$ are also learned from the same graph $(\mathbf{A}_f, \mathbf{X})$, their disparity should also be enhanced by HSIC:

$$HSIC(\mathbf{Z}_F, \mathbf{Z}_{CF}) = (n-1)^{-2}tr(\mathbf{R}\mathbf{K}_F\mathbf{R}\mathbf{K}_{CF}). \tag{3.14}$$

Then we set the disparity constraint as $\mathcal{L}_d$ where

$$\mathcal{L}_d = HSIC(\mathbf{Z}_T, \mathbf{Z}_{CT}) + HSIC(\mathbf{Z}_F, \mathbf{Z}_{CF}). \tag{3.15}$$

We use the output embedding $\mathbf{Z}$ in Eq. (3.10) for semi-supervised multi-class classification with a linear transformation and a softmax function. Denote the class predictions for n nodes as $\hat{\mathbf{Y}} = [\hat{y}_{ic}] \in \mathbb{R}^{n \times C}$ where $\hat{y}_{ic}$ is the probability of node i belonging to class c. Then $\hat{\mathbf{Y}}$ can be calculated as follows:

$$\hat{\mathbf{Y}} = softmax(\mathbf{W} \cdot \mathbf{Z} + \mathbf{b}), \tag{3.16}$$

where $softmax(x) = \frac{\exp(x)}{\Sigma_{c=1}^C exp(x_c)}$ is actually a normalizer across all classes.

Suppose the training set is L, for each $l \in L$ the real label is $\mathbf{Y}_l$ and the predicted label is $\hat{\mathbf{Y}}_l$. Then the cross-entropy loss for node classification over all training nodes is represented as $\mathcal{L}_t$ where

$$\mathcal{L}_t = -\sum_{l \in L} \sum_{i=1}^{C} \mathbf{Y}_l \ln\hat{\mathbf{Y}}_l. \tag{3.17}$$

Combining the node classification task and constraints, we have the following overall objective function:

$$\mathcal{L} = \mathcal{L}_t + \gamma\mathcal{L}_c + \beta\mathcal{L}_d, \tag{3.18}$$

where γ and β are parameters of the consistency and disparity constraint terms. With the guide of labeled data, we can optimize the proposed model via backpropagation and learn the embedding of nodes for classification.

The more detailed method description and experiment validation can be seen in [205].

3.2.4 Experiments

3.2.4.1 Experimental Setting

Our proposed AM-GCN is evaluated on six real-world datasets **Citeseer** [102], **UAI2010** [196], **ACM** [200], **BlogCatalog** [130], **Flickr** [130], and **CoraFull** [9]. In addition, we compare AM-GCN with two types of state-of-the-art methods, covering one network embedding algorithm, i.e., **DeepWalk** [154], and five graph neural network-based methods, i.e., **ChebNet, GCN** [102], k**NN-GCN, GAT** [189], and **MixHop** [2].

Table 3.1 Node classification results (%). (Bold: best; Underline: runner-up)

Datasets	Metrics	L/C	DW	ChebNet	GCN	kNN-GCN	GAT	MixHop	AM-GCN
Citeseer	ACC	20	43.47	69.80	70.30	61.35	<u>72.50</u>	71.40	**73.10**
		40	45.15	71.64	<u>73.10</u>	61.54	73.04	71.48	**74.70**
		60	48.86	73.26	74.48	62.38	<u>74.76</u>	72.16	**75.56**
	F1	20	38.09	65.92	67.50	58.86	<u>68.14</u>	66.96	**68.42**
		40	43.18	68.31	<u>69.70</u>	59.33	69.58	67.40	**69.81**
		60	48.01	70.31	<u>71.24</u>	60.07	**71.60**	69.31	70.92
UAI2010	ACC	20	42.02	50.02	49.88	<u>66.06</u>	56.92	61.56	**70.10**
		40	51.26	58.18	51.80	<u>68.74</u>	63.74	65.05	**73.14**
		60	54.37	59.82	54.40	<u>71.64</u>	68.44	67.66	**74.40**
	F1	20	32.93	33.65	32.86	<u>52.43</u>	39.61	49.19	**55.61**
		40	46.01	38.80	33.80	<u>54.45</u>	45.08	53.86	**64.88**
		60	44.43	40.60	34.12	54.78	48.97	<u>56.31</u>	**65.99**
ACM	ACC	20	62.69	75.24	<u>87.80</u>	78.52	87.36	81.08	**90.40**
		40	63.00	81.64	<u>89.06</u>	81.66	88.60	82.34	**90.76**
		60	67.03	85.43	<u>90.54</u>	82.00	90.40	83.09	**91.42**
	F1	20	62.11	74.86	<u>87.82</u>	78.14	87.44	81.40	**90.43**
		40	61.88	81.26	<u>89.00</u>	81.53	88.55	81.13	**90.66**
		60	66.99	85.26	<u>90.49</u>	81.95	90.39	82.24	**91.36**
Blog	ACC	20	38.67	38.08	69.84	<u>75.49</u>	64.08	65.46	**81.98**
		40	50.80	56.28	71.28	<u>80.84</u>	67.40	71.66	**84.94**
		60	55.02	70.06	72.66	<u>82.46</u>	69.95	77.44	**87.30**
	F1	20	34.96	33.39	68.73	<u>72.53</u>	63.38	64.89	**81.36**
		40	48.61	53.86	70.71	<u>80.16</u>	66.39	70.84	**84.32**
		60	53.56	68.37	71.80	<u>81.90</u>	69.08	76.38	**86.94**
Flickr	ACC	20	24.33	23.26	41.42	<u>69.28</u>	38.52	39.56	**75.26**
		40	28.79	35.10	45.48	<u>75.08</u>	38.44	55.19	**80.06**
		60	30.10	41.70	47.96	<u>77.94</u>	38.96	64.96	**82.10**
	F1	20	21.33	21.27	39.95	<u>70.33</u>	37.00	40.13	**74.63**
		40	26.90	33.53	43.27	<u>75.40</u>	36.94	56.25	**79.36**
		60	27.28	40.17	46.58	<u>77.97</u>	37.35	65.73	**81.81**
CoraFull	ACC	20	29.33	53.38	56.68	41.68	<u>58.44</u>	47.74	**58.90**
		40	36.23	58.22	60.60	44.80	<u>62.98</u>	57.20	**63.62**
		60	40.60	59.84	62.00	46.68	<u>64.38</u>	60.18	**65.36**
	F1	20	28.05	47.59	52.48	37.15	<u>54.44</u>	45.07	**54.74**
		40	33.29	53.47	55.57	40.42	<u>58.30</u>	53.55	**59.19**
		60	37.95	54.15	56.24	43.22	<u>59.61</u>	56.40	**61.32**

3.2.4.2 Node Classification

Node classification results are reported in Table 3.1, where L/C means the number of labels per class. We have the following observations:

(1) Compared with all baselines, the proposed AM-GCN generally achieves the best performance on all datasets with all label rates. Especially, for ACC, AM-GCN achieves maximum relative improvements of 8.59% on BlogCatalog and 8.63% on Flickr. The results demonstrate the effectiveness of AM-GCN.

(2) AM-GCN consistently outperforms GCN and kNN-GCN on all the datasets, indicating the effectiveness of the adaptive fusion mechanism in AM-GCN, because it can extract more useful information than only performing GCN and kNN-GCN, respectively.

(3) Comparing with GCN and kNN-GCN, we can learn that there does exist structural difference between topology graphs and feature graphs and performing GCN on traditional topology graphs does not always show better result than on feature graphs. For example, in BlogCatalog, Flickr, and UAI2010, the feature graphs perform better than topology. This further confirms the necessity of introducing feature graphs to GCN.

(4) Compared with GCN, the improvement of AM-GCN is more substantial on the datasets with better feature graph (kNN), such as UAI2010. This implies that AM-GCN introduces a better and more suitable kNN graph for label to supervise feature propagation and node representation learning.

3.3   Beyond Low-Frequency Information in Graph Convolutional Networks

3.3.1   Overview

In general, GNNs update node representations by aggregating information from neighbors, which can be seen as a special form of low-pass filter [117, 207]. Some recent studies [145, 212] show that the smoothness of signals, i.e., low-frequency information, are the key to the success of GNNs. However, is the low-frequency information all we need and what roles do other information play in GNNs? This is a fundamental question that motivates us to rethink whether GNNs comprehensively exploit the information in node features when learning node representation.

In order to verify whether the information of other frequencies is useful, we focus on low-frequency and high-frequency signals as an example, and present experiments to assess their roles. The results clearly show that both of them are helpful for learning node representations. Specifically, we find that when a network exhibits disassortativity, high-frequency signals perform much better than low-frequency signals. This implies that the high-frequency information, which is largely eliminated by the current GNNs, is not always useless, and the low-frequency information is not always optimal for the complex networks. Once the weakness of low-frequency information in GNNs is identified, a natural question is how to use signals of different frequencies in GNNs and, at the same time, make GNNs suitable for different types of networks.

To answer this question, two challenges need to be solved: (1) Both the low-frequency and high-frequency signals are the parts of the raw features. Traditional filter is specifically designed for one certain signal, and cannot well extract different frequency signals simultaneously. (2) Even we can extract different information, however, the assortativity of real-world networks is usually agnostic and varies greatly, moreover, the correlation between task and different information is very complex, so it is difficult to decide what kind of signals should be used: raw features, low-frequency signals, high-frequency signals, or their combination. We design a general frequency adaptation graph convolutional network called FAGCN, to adaptively aggregate different signals from neighbors or itself. We first employ the theory of graph signal processing to formally define an enhanced low-pass and high-pass filter to separate the low-frequency and high-frequency signals from the raw features. Then we design a self-gating mechanism to adaptively integrate the low-frequency signals, high-frequency signals, and raw features, without knowing the assortativity of network. Extensive experiments on six real-world networks validate that FAGCN has advantages over state of the arts.

3.3.2 Investigation

Taking the low-frequency and high-frequency signals as an example, we analyze their roles in learning node representations. Specifically, we test their performance of node classification on a series of synthetic networks. The main idea is to gradually increase the disassortativity of the synthetic networks, and observe how the performance of these two signals changes. We generate a network with 200 nodes and randomly divide them into 2 classes. For each node in class one, we sample a 20-dimensional feature vector from Gaussian distribution $\mathcal{N}(0.5, 1)$, while for the nodes in class two, the distribution is $\mathcal{N}(-0.5, 1)$. Besides, the connections in the same class are generated from a Bernoulli distribution with probability $p = 0.05$, and the probability of connections between two classes q varies from 0.01 to 0.1. When q is small, the network exhibits assortativity; as q increases, the network gradually exhibits disassortativity. We then apply the low-pass and high-pass filters, described in Sect. 3.3.3.1, to node classification task. Half of the nodes are used for training and the remains are used for testing.

Figure 3.2a illustrates that with the increase of inter-connection q, the accuracy of low-frequency signals decreases, while the accuracy of high-frequency signals increases gradually. This proves that both the low-frequency and high-frequency signals are helpful in learning node representations. The reason why existing GNNs fail when q increases is that, as shown in Fig. 3.2b, they only aggregate low-frequency signals from neighbors, i.e., making the node representations become similar, regardless of whether nodes belong to the same class, thereby losing the discrimination. When the network becomes disassortative, the effec-

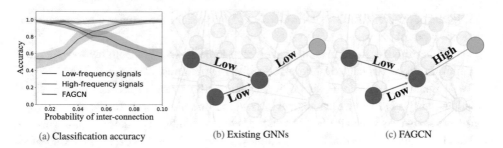

(a) Classification accuracy   (b) Existing GNNs   (c) FAGCN

Fig. 3.2 a Classification accuracy of low-frequency signals, high-frequency signals, and our model FAGCN. X-axis denotes probability of inter-connection q. **b** Existing GNNs aggregate the low-frequency signals of neighbors. **c** FAGCN aggregates the low-frequency signals of neighbors within the same class and high-frequency signals of neighbors from different classes, where the color indicates the node label

tiveness of high-frequency signals appears, but as shown in Fig. 3.2a, a single filter cannot achieve optimal results in all cases. Our proposed FAGCN, which combines the advantages of both low-pass and high-pass filters, can aggregate the low-frequency signals within the same class and high-frequency signals from different classes, as shown in Fig. 3.2c, thereby showing the best performance on every synthetic network.

3.3.3   The FAGCN Method

Consider an undirected graph $G = (V, E)$ with adjacency matrix $A \in \mathbb{R}^{N \times N}$, where V is a set of nodes with $|V| = N$ and E is a set of edges. The normalized graph Laplacian matrix is defined as $L = I_n - D^{-1/2} A D^{-1/2}$, where $D \in \mathbb{R}^{N \times N}$ is a diagonal degree matrix with $D_{i,i} = \sum_j A_{i,j}$ and I_n denotes the identity matrix. Because L is a real symmetric matrix, it has a complete set of orthonormal eigenvectors $\{u_l\}_{l=1}^n \in \mathbb{R}^n$, each of which has a corresponding eigenvalue $\lambda_l \in [0, 2]$ [31]. Through the eigenvalues and eigenvectors, we have $L = U \Lambda U^\top$, where $\Lambda = diag([\lambda_1, \lambda_2, \ldots, \lambda_n])$.

Graph Fourier Transform. According to theory of graph signal processing [170], we can treat the eigenvectors of normalized Laplacian matrix as the bases in graph Fourier transform. Given a signal $x \in \mathbb{R}^n$, the graph Fourier transform is defined as $\hat{x} = U^\top x$, and the inverse graph Fourier transform is $x = U \hat{x}$. Thus, the convolutional $*_G$ between the signal x and convolution kernel f is: $f *_G x = U \left((U^\top f) \odot (U^\top x) \right) = U g_\theta U^\top x$, where $\odot$ denotes the element-wise product of vectors and g_θ is a diagonal matrix, which represents the convolutional kernel in the spectral domain, replacing $U^\top f$. Spectral CNN [12] uses a non-parametric convolutional kernel $g_\theta = diag(\{\theta_i\}_{i=1}^n)$. ChebyNet [37] parameterizes convolutional kernel with a polynomial expansion $g_\theta = \sum_{k=0}^{K-1} \alpha_k \Lambda^k$. GCN defines the convolutional kernel as $g_\theta = I - \Lambda$.

3.3.3.1 Separation

As discussed in the aforementioned section, both the low-frequency and high-frequency signals are helpful for learning node representations. To make full use of them, we design a low-pass filter $\mathcal{F}_L$ and a high-pass filter $\mathcal{F}_H$ to separate the low-frequency and high-frequency signals from the node features:

$$\mathcal{F}_L = \varepsilon I + D^{-1/2} A D^{-1/2} = (\varepsilon + 1)I - L,$$
$$\mathcal{F}_H = \varepsilon I - D^{-1/2} A D^{-1/2} = (\varepsilon - 1)I + L, \tag{3.19}$$

where ε is a scaling hyper-parameter limited in $[0, 1]$. If we use $\mathcal{F}_L$ and $\mathcal{F}_H$ to replace the convolutional kernel f. The signal x is filtered by $\mathcal{F}_L$ and $\mathcal{F}_H$ as

$$\mathcal{F}_L *_G x = U[(\varepsilon + 1)I - \Lambda]U^\top x = \mathcal{F}_L \cdot x,$$
$$\mathcal{F}_H *_G x = U[(\varepsilon - 1)I + \Lambda]U^\top x = \mathcal{F}_H \cdot x. \tag{3.20}$$

Therefore, the convolutional kernel of $\mathcal{F}_L$ is $g_\theta = (\varepsilon + 1)I - \Lambda$, rewritten as $g_\theta(\lambda_i) = \varepsilon + 1 - \lambda_i$. When $\lambda_i > 1 + \varepsilon$, $g_\theta(\lambda_i) < 0$, which gives a negative amplitude. To avoid this, we consider the second-order convolution kernel $\mathcal{F}_L^2$ with $g_\theta(\lambda_i) = (\varepsilon + 1 - \lambda_i)^2$. When $\lambda_i = 0$, $g_\theta(\lambda_i) = (\varepsilon + 1)^2 > 1$ and when $\lambda_i = 2$, $g_\theta(\lambda_i) = (\varepsilon - 1)^2 < 1$, which amplifies the low-frequency signals and restrains the high-frequency signals.

Separating the low-frequency and high-frequency signals from the node features provides a feasible way to deal with different networks, e.g., low-frequency signals for assortative networks and high-frequency signals for disassortative networks. However, this way has two disadvantages: one is that selecting signals requires a priori knowledge, i.e., we actually do not know whether a network is assortative or disassortative beforehand. The other is that, as in Eq. 3.20, it requires matrix multiplication, which is undesirable for large graphs [74]. Therefore, an efficient method that can adaptively aggregate low-frequency and high-frequency signals is desired.

3.3.3.2 Aggregation

The inputs of our model are the node features, $\mathbf{H} = \{\mathbf{h}_1, \mathbf{h}_2, \ldots, \mathbf{h}_N\} \in \mathbb{R}^{N \times F}$, where F is the dimension of the node features. For the purpose of frequency adaptation, a basic idea is to use the attention mechanism to learn the proportion of low-frequency and high-frequency signals:

$$\tilde{\mathbf{h}}_i = \alpha_{ij}^L (\mathcal{F}_L \cdot \mathbf{H})_i + \alpha_{ij}^H (\mathcal{F}_H \cdot \mathbf{H})_i = \varepsilon \mathbf{h}_i + \sum_{j \in \mathcal{N}_i} \frac{\alpha_{ij}^L - \alpha_{ij}^H}{\sqrt{d_i d_j}} \mathbf{h}_j, \tag{3.21}$$

where $\tilde{\mathbf{h}}_i$ is the aggregated representation of node i. $\mathcal{N}_i$ and d_i denote the neighbor set and degree of node i, respectively. α_{ij}^L and α_{ij}^H represent the proportions of node j's low-frequency and high-frequency signals to node i. We set $\alpha_{ij}^L + \alpha_{ij}^H = 1$ and $\alpha_{ij}^G = \alpha_{ij}^L - \alpha_{ij}^H$.

In order to learn the coefficients α_{ij}^G effectively, we need to consider the features of both the node itself and its neighbors. Therefore, we propose a shared *self-gating* mechanism $\mathbb{R}^F \times \mathbb{R}^F \to \mathbb{R}$ to learn the coefficients:

$$\alpha_{ij}^G = \tanh \left(\mathbf{g}^\top \left[\mathbf{h}_i \parallel \mathbf{h}_j \right] \right), \tag{3.22}$$

where $\parallel$ denotes the concatenation operation, $\mathbf{g} \in \mathbb{R}^{2F}$ can be seen as a shared convolutional kernel [189], and $\tanh(\cdot)$ is the hyperbolic tangent function, which can naturally limit the value of α_{ij}^G in $[-1, 1]$. Besides, to make use of the structural information, we only calculate the coefficients among the node and its first-order neighbors $\mathcal{N}_i$.

After calculating α_{ij}^G, we can aggregate the representations of neighbors:

$$\mathbf{h}_i^{'} = \varepsilon \mathbf{h}_i + \sum_{j \in \mathcal{N}_i} \frac{\alpha_{ij}^G}{\sqrt{d_i d_j}} \mathbf{h}_j, \tag{3.23}$$

where $\mathbf{h}_i^{'}$ denotes the aggregated representation of node i. Note that when aggregating information from neighbors, the degrees are used to normalize the coefficients, thus preventing the aggregated representations from being too large.

The more detailed method description and experiment validation can be seen in [8].

3.3.4   Experiments

3.3.4.1 Experimental Settings

Assortative datasets. We choose the commonly used *citation networks*, e.g., Cora, Citeseer, and Pubmed for assortative datasets. Edges in these networks represent the citation relationship between two papers (undirected), node features are the bag-of-words vector of the papers and labels are the fields of papers. In each network, we use 20 labeled nodes per class for training, 500 nodes for validation, and 1000 nodes for testing. Details can be found in [102].

Disassortative datasets. We consider the *Wikipedia networks*[1] and *Actor co-occurrence network* [182] for disassortative datasets. Chameleon and Squirrel are two Wikipedia networks. Edges represent the hyperlinks between two pages, node features are some informative nouns in the pages, and labels correspond to the traffic of the pages. In Actor co-occurrence network, each node represents an actor, and the edges denote the collaborations of them. Node features are the keywords in Wikipedia and labels are the types of actors.

---

[1] http://snap.stanford.edu/data/wikipedia-article-networks.html.

Table 3.2 Summary of node classification results (in percent)

Method	Cora (%)	Citeseer (%)	Pubmed (%)
SGC	81.0	71.9	78.9
GCN	81.5	70.3	79.0
GWNN	82.8	71.7	79.1
ChebNet	81.2	69.8	74.4
GraphHeat	83.7	72.5	**80.5**
GIN	77.6	66.1	77.0
GAT	83.0	72.5	79.0
MoNet	81.7	–	78.8
APPNP	83.7	72.1	79.2
GraphSAGE	82.3	71.2	78.5
FAGCN	**84.1±0.5**	**72.7±0.8**	79.4±0.3

3.3.4.2 Node Classification Results

Classification on Assortative Graphs. The performance of different methods on assortative networks is summarized in Table 3.2. GraphHeat designs a low-pass filer through heat kernel, which can better capture the low-frequency information than GCN [212]. Therefore, it performs best in the baselines. But we can see that FAGCN exceed the benchmarks on most networks due to the enhanced low-pass filter, which validates the importance of low-pass filters in the assortative networks.

Classification on Disassortative Graphs. Besides, the performance on disassortative networks is illustrated in Fig. 3.3. Note that we do not choose all baselines because the methods focusing on low-pass filtering have poor performance, and we use GCN and GAT as representatives. In addition, APPNP leverages residual connection to preserve the information of raw features, ChebyNet uses Chebyshev polynomials to approximate arbitrary convolution kernels and Geom-GCN is the state of the art on disassortative networks. Therefore, com-

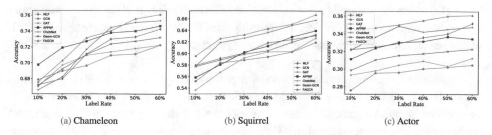

(a) Chameleon   (b) Squirrel   (c) Actor

Fig. 3.3 Classification accuracy of different methods under different label rates on disassortative networks

paring FAGCN with these baselines can reflect the superiority of FAGCN. From Fig. 3.3, we can see that GCN and GAT perform worse than other methods, which indicates that only using low-pass filters is not suitable for disassortative networks. APPNP and ChebNet perform better than GCN and GAT, which shows that the raw features and polynomials can preserve the high-frequency information to some extent. Finally, FAGCN performs best in most datasets and label rates, which reflects the superiority of our method.

3.4   Graph Structure Estimation Neural Networks

3.4.1   Overview

Although existing GNNs have been successfully applied in a wide variety of scenarios, they rely on one fundamental assumption that the observed topology is ground-truth information, so the messages can be well spread within the corresponding community. But in fact, as graphs are usually extracted from complex interaction systems, such assumption could always be violated. In order to preserve the essential information and eliminate the noise in message passing, it is imperative to explore an optimal message-passing structure, i.e., graph topology, for GNNs.

Nevertheless, it is technically challenging to effectively learn an optimal graph structure for GNNs. Particularly, two obstacles need to be addressed. (1) The graph generation mechanism should be taken into consideration. It is well established in network science literature [138] that the graph generation is potentially governed by some underlying principles, e.g., the configuration model [137]. Considering these principles fundamentally drives the learned graph to maintain a regular global structure and be more robust to noise in real observations. Unfortunately, majority of current methods parameterize each edge locally [27, 52, 94] and do not account for the underlying generation of graph, so that the resulted graph has a lower tolerance for noise and sparsity. (2) Multifaceted information should be injected to reduce bias. Learning graph structure from one information source inevitably leads to bias and uncertainty. It makes sense that the confidence of an edge would be greater if this edge exists under multiple measurements. Thus, a reliable graph structure ought to make allowance for comprehensive information, although it is complicated to obtain multi-view measurements and depict their relationship with GNNs. Existing approaches [95, 241] utilize the feature similarity, making the learned graph more susceptible to the bias of single view.

To address the aforementioned issues, we propose Graph structure Estimation neural Networks (GEN) to improve the node classification performance through estimating an appropriate graph structure for GNNs. We firstly analyze the properties of GNNs to match proper graph generation mechanism. GNNs, as low-pass filters [7, 116, 207] which smooth neighborhood to make the representations of proximal nodes similar, are suitable to graphs with community structure [62]. Therefore, we attach a structure model to the graph generation, hypothesizing that the estimated graph is drawn from Stochastic Block Model (SBM) [81]. Furthermore, in addition to the observed graph and node feature, we creatively inject multi-order neighborhood information to circumvent bias and present an observation model to jointly treat above multi-view information as observations of the optimal graph. In order to estimate the optimal graph, we construct observations during GNN training, then apply Bayesian inference based on structure and observation models to infer the entire posterior distribution over graph structure. Finally, the estimated graph and the parameters of GNNs achieve mutual, positive reinforcement through elaborately iterative optimization.

3.4.2   The GEN Method

Let $\mathcal{G} = (\mathcal{V}, \mathcal{E}, \mathbf{X})$ be a graph, where $\mathcal{V}$ is the set of N nodes $\{v_1, v_2, \ldots, v_N\}$, $\mathcal{E}$ is the set of edges, $\mathbf{X} = [\mathbf{x}_1, \mathbf{x}_2, \ldots, \mathbf{x}_N] \in \mathbb{R}^{N \times D}$ represents the node feature matrix, and $\mathbf{x}_i$ is the feature vector of node v_i. The edges describe the relations between nodes and can be represented by an adjacency matrix $\mathbf{A} \in \mathbb{R}^{N \times N}$, where A_{ij} denotes the relation between nodes v_i and v_j. Following the common semi-supervised node classification setting, only small part of nodes $\mathcal{V}_L = \{v_1, v_2, \ldots, v_l\}$ are associated with corresponding labels $\mathcal{Y}_L = \{y_1, y_2, \ldots, y_l\}$, where y_i is label of v_i.

Given graph $\mathcal{G} = (\mathcal{V}, \mathcal{E}, \mathbf{X})$ and the partial labels $\mathcal{Y}_L$, the goal of graph structure learning for GNNs is to simultaneously learn an optimal adjacency matrix $\mathbf{S} \in \mathcal{S} = [0, 1]^{N \times N}$ and the GNN parameters Θ to improve node classification performance for unlabeled nodes. The objective function can be formulated as

$$\min_{\Theta, \mathbf{S}} \mathcal{L}(\mathbf{A}, \mathbf{X}, \mathcal{Y}_L) = \sum_{v_i \in \mathcal{V}_L} \ell\left(f_\Theta(\mathbf{X}, \mathbf{S})_i, y_i\right), \qquad (3.24)$$

where $f_\Theta : \mathcal{V}_L \to \mathcal{Y}_L$ is the function learned by GNNs, $f_\Theta(\mathbf{X}, \mathbf{S})_i$ is the prediction of node v_i, and $\ell(\cdot, \cdot)$ is to measure the difference between prediction and true label, such as cross entropy.

3.4.2.1 Observation Construction

Without loss of generality, we choose representative GCN as backbone. To begin with, we feed the original graph $\mathcal{G} = (\mathcal{V}, \mathcal{E}, \mathbf{X})$ into the vanilla GCN to construct an initial observation set O for subsequent graph estimation.

Specifically, GCN follows a neighborhood aggregation strategy, which iteratively updates the presentation of a node by aggregating representations of its neighbors. Formally, the kth-layer aggregation rule of GCN is

$$\mathbf{H}^{(k)} = \sigma\left(\tilde{\mathbf{D}}^{-\frac{1}{2}} \tilde{\mathbf{A}} \tilde{\mathbf{D}}^{-\frac{1}{2}} \mathbf{H}^{(k-1)} \mathbf{W}^{(k)}\right). \qquad (3.25)$$

Here, $\tilde{\mathbf{A}}$ is the normalized adjacency matrix and $\tilde{D}_{ii} = \sum_j \tilde{A}_{ij}$. $\mathbf{W}^{(k)}$ is a layer-wise train-able weight matrix, and σ denotes an activation function. $\mathbf{H}^{(k)} \in \mathbb{R}^{N \times d}$ is the matrix of node representations in the kth layer, and $\mathbf{H}^{(0)} = \mathbf{X}$. In terms of l-layer GCN, the activation function of the last layer l is row-wise softmax and predictions $\mathbf{Z} = \mathbf{H}^{(l)}$. The GCN parameters $\Theta = (\mathbf{W}^{(1)}, \mathbf{W}^{(2)}, \ldots, \mathbf{W}^{(l)})$ can be trained via gradient descent.

The current GCN acts directly on the observed graph $\mathbf{A}$ which is extracted from the real-world complex systems and is usually noisy. To estimate an optimal graph structure for GCN, we need to construct multifaceted observations that could be ensembled to resist bias. Fortunately, after k iterations of aggregation, node representation captures the structural information within its k-order graph neighborhood which provides local to global information. On the other hand, node pairs with similar neighborhood are possibly far away in the graph but apt to the same communities. If these informative node pairs are employed, they could provide useful clues for downstream classification. Therefore, we attempt to connect these distant but similar nodes in our estimated graph to enhance the performance of GCN.

Specifically, we fix the GCN parameters Θ and take out the node representations $\mathcal{H} = \{\mathbf{H}^{(0)}, \mathbf{H}^{(1)}, \ldots, \mathbf{H}^{(l)}\}$ to construct kNN graphs $\{\mathbf{O}^{(0)}, \mathbf{O}^{(1)}, \ldots, \mathbf{O}^{(l)}\}$ as observations of the optimal graph, where $\mathbf{O}^{(i)}$ is the adjacency matrix of kNN graph generated by $\mathbf{H}^{(i)}$ and characterizes the similarity of i-order neighborhood. Obviously, the original graph $\mathbf{A}$ is also an important external observation of the optimal graph, thus we combine it with kNN graphs to form the complete observation set $O = \{\mathbf{A}, \mathbf{O}^{(0)}, \mathbf{O}^{(1)}, \ldots, \mathbf{O}^{(l)}\}$. These observations reflect the optimal graph structure from varied views and can be ensembled to infer a more reliable graph structure.

As preliminary preparations, these observations O, predictions $\mathbf{Z}$, and labels $\mathcal{Y}_L$ will be fed into estimator to accurately infer the posterior distribution of the graph structure. In the following, we will introduce the inference in detail.

Graph Estimator. Until now, the question we would like to answer is: given these available observations O, what is the best estimated graph for GCN? These observations reveal the optimal graph structure from different perspectives, but they may be unreliable or incomplete, and we do not have a priori that how accurate any of the information is. Under this daunting circumstance, it is not straightforward to answer this question directly, but it is relatively easy to answer the reverse question. Imagining that a graph with community structure has been generated, we could calculate the probability of mapping this graph to these observations. If we can do this, Bayesian inference allows us to invert the calculation and compute the posterior distribution of graph structure, hence our goal is achieved. The procedure is formalized as follows.

3.4.2.2 Structure Model

Let us denote the optimal graph which we are trying to estimate, as a symmetric adjacency matrix $\mathbf{G}$, and we firstly propose a structure model to represent the underlying structure generation of optimal graph $\mathbf{G}$.

Considering the local smoothing nature of GCN, a good choice is Stochastic Block Model (SBM) which is widely used for community detection and applicable to model a graph that has relatively strong community structure [98, 160]. The values of within- and between-community parameters in the fitted block model can constrain the homophily of the estimate. Although there are also other SBM variants, e.g., degree-corrected SBM [98]. But in this section, the effectiveness of vanilla SBM as structure model has been verified. We leave more complex structure models as future work and believe it will further improve performance.

Particularly, this process of generating optimal graph $\mathbf{G}$ takes the form of a probability distribution $P(\mathbf{G}|\mathbf{\Omega}, \mathbf{Z}, \mathcal{Y}_{\mathbf{L}})$. Here, $\mathbf{\Omega}$ represents the parameters of SBM which assumes that the probability of an edge between nodes depends only on their communities. For instance, the probability of an edge between node v_i with community c_i and node v_j with community c_j is $\Omega_{c_i c_j}$. Therefore, $\mathbf{\Omega}$ indicates the probabilities of within- and between-community connections. Given parameters $\mathbf{\Omega}$, predictions $\mathbf{Z}$, and labels $\mathcal{Y}_L$, the probability of generating graph $\mathbf{G}$ is formalized as

$$P(\mathbf{G}|\mathbf{\Omega}, \mathbf{Z}, \mathcal{Y}_L) = \prod_{i<j} \Omega_{c_i c_j}^{G_{ij}} (1 - \Omega_{c_i c_j})^{1-G_{ij}}, \tag{3.26}$$

where

$$c_i = \begin{cases} y_i & \text{if } v_i \in \mathcal{V}_L, \\ z_i & \text{otherwise}, \end{cases} \tag{3.27}$$

which means that generating an edge between node v_i and v_j in optimal graph $\mathbf{G}$ depends only on the probability $\Omega_{c_i c_j}$ related to community identifications c_i and c_j. Here, in order to obtain more accurate community identifications, we use the label-corrected predictions that replace the community identifications of nodes in training set directly with labels.

3.4.2.3 Observation Model

Please note that the structure model represents our prior knowledge or constrain about the underlying structure before observing any data. In fact, in what form the optimal graph structure exists is a mystery, and the things that can be done are to combine its external observations to infer it.

Therefore, we introduce an observation model to describe how the optimal graph $\mathbf{G}$ maps onto observations, which assumes that the observations of edges are independent identically distributed Bernoulli random variables, conditional on the presence or absence of an edge in the optimal graph. This assumption is well accepted in literature, e.g., community detection [136] and graph generation [171, 227], which has been proven to be feasible.

$P(O|\mathbf{G}, \alpha, \beta)$ is the probability of these observations O given the optimal graph $\mathbf{G}$ and model parameters α and β. Specifically, we parameterize the possible observations by two probabilities: the true-positive rate α, which is the probability of observing an edge where one truly exists in the optimal graph $\mathbf{G}$, and the false-positive rate β, which is the probability of observing an edge where none exists in the optimal graph $\mathbf{G}$. The remaining possibilities of true-negative and false-negative rates are $1 - \beta$ and $1 - \alpha$, respectively. Let us suppose that out of the M (i.e., $|O|$) observations, we observe an edge on E_{ij} of them, and no edge on the remaining $M - E_{ij}$. Plugging these definitions, we can write the specific form of $P(O|\mathbf{G}, \alpha, \beta)$ as

$$
P(O|\mathbf{G}, \alpha, \beta) = \prod_{i<j} \left[\alpha^{E_{ij}}(1-\alpha)^{M-E_{ij}} \right]^{G_{ij}} \times \left[\beta^{E_{ij}}(1-\beta)^{M-E_{ij}} \right]^{1-G_{ij}}. \tag{3.28}
$$

If there is truly an edge in optimal graph $\mathbf{G}$, the probability of observing E_{ij} edges between node v_i and v_j out of total M observations is succinctly written as $\alpha^{E_{ij}}(1-\alpha)^{M-E_{ij}}$. If not, the probability is $\beta^{E_{ij}}(1-\beta)^{M-E_{ij}}$.

Graph Estimation. Having clearly clarified our models for structure and observation, we now present our graph estimation process based on Bayesian inference.

It is difficult to calculate posterior probability $P(\mathbf{G}, \mathbf{\Omega}, \alpha, \beta|O, \mathbf{Z}, \mathcal{Y}_L)$ for optimal graph directly. However, combining our above models and applying Bayes rule, we then have

$$
P(\mathbf{G}, \mathbf{\Omega}, \alpha, \beta|O, \mathbf{Z}, \mathcal{Y}_L) = \frac{P(O|\mathbf{G}, \alpha, \beta) P(\mathbf{G}|\mathbf{\Omega}, \mathbf{Z}, \mathcal{Y}_L) P(\mathbf{\Omega}) P(\alpha) P(\beta)}{P(O, \mathbf{Z}, \mathcal{Y}_L)}, \tag{3.29}
$$

where $P(\mathbf{\Omega})$, $P(\alpha)$, $P(\beta)$, and $P(O, \mathbf{Z}, \mathcal{Y}_L)$ are the probabilities of the parameters and available data, which we assume to be independent.

We get an expression for the posterior probability of the parameters $\mathbf{\Omega}$, α and β, by summing all possible values of optimal graph $\mathbf{G}$:

$$
P(\mathbf{\Omega}, \alpha, \beta|O, \mathbf{Z}, \mathcal{Y}_L) = \sum_{\mathbf{G}} P(\mathbf{G}, \mathbf{\Omega}, \alpha, \beta|O, \mathbf{Z}, \mathcal{Y}_L). \tag{3.30}
$$

Maximizing this posterior probability *w.r.t.* $\mathbf{\Omega}$, α and β will give maximum a posteriori (MAP) estimates for these parameters. Based on these MAP estimates, we can calculate the estimated adjacency matrix $\mathbf{Q}$ for optimal graph $\mathbf{G}$

$$
Q_{ij} = \sum_{\mathbf{G}} q(\mathbf{G}) G_{ij}, \tag{3.31}
$$

where quantity Q_{ij} is the posterior probability that there is an edge between nodes v_i and v_j, representing our confidence about whether the edge exists.

3.4.2.4 Iterative Optimization

Jointly optimizing GCN parameters Θ and estimated adjacency matrix $\mathbf{Q}$ is challenging. And the dependence between them exacerbates the difficulty. In this work, we use an alternating optimization schema to iteratively update Θ and $\mathbf{Q}$.

Update Θ. For semi-supervised node classification, we evaluate the cross-entropy error over all labeled examples $\mathcal{Y}_L$

$$\min_{\Theta} \mathcal{L}(\mathbf{A}, \mathbf{X}, \mathcal{Y}_L) = - \sum_{v_i \in \mathcal{V}_L} \mathbf{y}_i \ln \mathbf{z}_i, \tag{3.32}$$

which is a typical GCN optimization problem and we can learn parameters Θ via stochastic gradient descent.

Update Q. To update the estimated adjacency matrix $\mathbf{Q}$, we maximize Eq. (3.30) with Expectation-Maximization (EM) algorithm [38, 129, 136].

E-step. Since it is convenient to maximize not the probability itself but its logarithm, we apply Jensen's inequality to the log of Eq. (3.30)

$$\log P(\mathbf{\Omega}, \alpha, \beta | \mathbf{O}, \mathbf{Z}, \mathcal{Y}_L) \geq \sum_{\mathbf{G}} q(\mathbf{G}) \log \frac{P(\mathbf{G}, \mathbf{\Omega}, \alpha, \beta | \mathbf{O}, \mathbf{Z}, \mathcal{Y}_L)}{q(\mathbf{G})}, \tag{3.33}$$

where $q(\mathbf{G})$ is any non-negative function of $\mathbf{G}$ satisfying $\sum_{\mathbf{G}} q(\mathbf{G}) = 1$, which can be seen as a probability distribution over $\mathbf{G}$.

The right-hand side of the inequality Eq. (3.33) is maximized when the exact equality is achieved

$$q(\mathbf{G}) = \frac{P(\mathbf{G}, \mathbf{\Omega}, \alpha, \beta | \mathbf{O}, \mathbf{Z}, \mathcal{Y}_L)}{\sum_{\mathbf{G}} P(\mathbf{G}, \mathbf{\Omega}, \alpha, \beta | \mathbf{O}, \mathbf{Z}, \mathcal{Y}_L)}. \tag{3.34}$$

Substituting Eqs. (3.26) and (3.28) into Eq. (3.34), and eliminating the constants in fraction, we find the following expression for $q(\mathbf{G})$:

$$\begin{aligned} q(\mathbf{G}) &= \frac{\prod_{i<j} \left[\Omega_{c_i c_j} \alpha^{E_{ij}} (1-\alpha)^{M-E_{ij}} \right]^{G_{ij}} \left[(1-\Omega_{c_i c_j}) \beta^{E_{ij}} (1-\beta)^{M-E_{ij}} \right]^{1-G_{ij}}}{\sum_{\mathbf{G}} \prod_{i<j} \left[\Omega_{c_i c_j} \alpha^{E_{ij}} (1-\alpha)^{M-E_{ij}} \right]^{G_{ij}} \left[(1-\Omega_{c_i c_j}) \beta^{E_{ij}} (1-\beta)^{M-E_{ij}} \right]^{1-G_{ij}}} \\ &= \prod_{i<j} \frac{\left[\Omega_{c_i c_j} \alpha^{E_{ij}} (1-\alpha)^{M-E_{ij}} \right]^{G_{ij}} \left[(1-\Omega_{c_i c_j}) \beta^{E_{ij}} (1-\beta)^{M-E_{ij}} \right]^{1-G_{ij}}}{\Omega_{c_i c_j} \alpha^{E_{ij}} (1-\alpha)^{M-E_{ij}} + (1-\Omega_{c_i c_j}) \beta^{E_{ij}} (1-\beta)^{M-E_{ij}}}. \end{aligned} \tag{3.35}$$

Then further maximizing the right-hand side of Eq. (3.33) will give us the MAP estimates.

M-step. We can find the maximum over the parameters by differentiating. Taking derivatives of the right-hand side of Eq. (3.33) while holding $q(\mathbf{G})$ constant, and assuming that the priors are uniform, we have

$$\sum_{\mathbf{G}} q(\mathbf{G}) \sum_{i<j} \left[\frac{G_{ij}}{\Omega_{c_i c_j}} - \frac{1-G_{ij}}{1-\Omega_{c_i c_j}} \right] = 0, \tag{3.36}$$

$$\sum_{\mathbf{G}} q(\mathbf{G}) \sum_{i<j} G_{ij} \left[\frac{E_{ij}}{\alpha} - \frac{M-E_{ij}}{1-\alpha} \right] = 0, \tag{3.37}$$

$$\sum_{\mathbf{G}} q(\mathbf{G}) \sum_{i<j} \left(1 - G_{ij}\right) \left[\frac{E_{ij}}{\beta} - \frac{M - E_{ij}}{1 - \beta} \right] = 0. \tag{3.38}$$

The solution of these equations gives us MAP estimates for $\boldsymbol{\Omega}$, α, and β. Note that Eq. (3.36) depends only on the SBM and its solution gives the parameter values for structure model. Similarly, Eqs. (3.37) and (3.38) depend only on the observation model. For specific calculations, we swap the order of the summations and find that

$$\Omega_{rs} = \begin{cases} \frac{M_{rs}}{n_r n_s} & \text{if } r \neq s, \\ \frac{2M_{rr}}{n_r(n_r-1)} & \text{otherwise,} \end{cases} \tag{3.39}$$

where $n_r = \sum_i \delta_{c_i,r}$ and $M_{rs} = \sum_{i<j} Q_{ij} \delta_{c_i,r} \delta_{c_j,s}$. Thus Eq. (3.39) has the simple interpretation that the probability Ω_{rs} of an edge between community r and s is the average probabilities of the individual edges between all nodes in these two communities. Similar calculations are done for α and β

$$\alpha = \frac{\sum_{i<j} Q_{ij} E_{ij}}{M \sum_{i<j} Q_{ij}}, \tag{3.40}$$

$$\beta = \frac{\sum_{i<j} \left(1 - Q_{ij}\right) E_{ij}}{M \sum_{i<j} \left(1 - Q_{ij}\right)}. \tag{3.41}$$

To calculate the value of Q_{ij}, we substitute Eq. (3.35) into Eq. (3.31):

$$Q_{ij} = \frac{\Omega_{c_i c_j} \alpha^{E_{ij}} (1 - \alpha)^{M - E_{ij}}}{\Omega_{c_i c_j} \alpha^{E_{ij}} (1 - \alpha)^{M - E_{ij}} + (1 - \Omega_{c_i c_j}) \beta^{E_{ij}} (1 - \beta)^{M - E_{ij}}}. \tag{3.42}$$

The posterior distribution $q(\mathbf{G})$ can be conveniently rewritten in terms of Q_{ij} as

$$q(\mathbf{G}) = \prod_{i<j} Q_{ij}^{G_{ij}} \left(1 - Q_{ij}\right)^{1 - G_{ij}}. \tag{3.43}$$

To put that another way, the probability distribution over optimal graph is the product of independent Bernoulli distributions of the individual edges, with Bernoulli parameters Q_{ij}, which capture both the graph structure itself and the uncertainty in that structure. This leads to a natural EM algorithm for determining the values of the parameters and posterior distribution over possible graph structures. We perform the E-step by maximizing first over $q(\mathbf{G})$ with the parameters held constant; then go to the M-step over parameters $\boldsymbol{\Omega}$, α, and β with $q(\mathbf{G})$ held constant, and repeat these iterations until convergence.

The more detailed method description and experiment validation can be seen in [194].

3.4.3   Experiments

3.4.3.1 Experimental Settings

We validate the proposed GEN on six open graph datasets. Citation networks [102]: Cora, Citeseer, and Pubmed are benchmark citation network datasets; Wikipedia networks [151]: Chameleon and Squirrel are two page-page networks in Wikipedia with specific topics; and Actor co-occurrence network [151]: this dataset is the actor-only induced subgraph of the film-director-actor-writer network.

To evaluate the effectiveness of GEN, we compare it with three categories of representative GNNs, including two spectral-based methods (i.e., GCN [102] and ChebyNet [37]), two spatial-based methods (i.e., GAT [189] and GraphSAGE [74]), and two graph structure learning-based methods (i.e., LDS [52] and ProGNN [95]).

3.4.3.2 Node Classification Results

We evaluate the semi-supervised node classification performance of GEN against state-of-the-art baselines. In addition to the 20 labels per class training explored in previous work, we also evaluate the performance under more severely limited data scenarios where only 10 or 5 labels per class are available. In 5 and 10 labels per class cases, we construct the training set by using the first 5 or 10 labels in the original partition, while keeping the validation and testing sets unchanged. In Table 3.3, we report the mean and standard deviation results over 10 independent trials with different random seeds. Based on the results, we have

(1) GEN consistently outperforms other baselines on six datasets, especially under reduced-label and disassortative settings, which demonstrates that our ingeniously designed graph estimation framework can boost node classification performance in a robust manner. We find that as the label rate and homophily decrease, the performance of GNNs drops quickly and the improvement of GEN becomes more pronounced. These phenomena are in line with our expectations that noisy or sparse observed graphs prevent GNNs from aggregating effective information, and our estimated graphs alleviate this issue.

(2) The overwhelming performance superiority of GEN over backbone GCN implies that GEN is capable of estimating suitable structure, so that graph structure estimation and parameter optimization of GCN reinforce each other.

(3) In comparison with other graph structure learning-based methods, our performance improvement illustrates that explicitly constraining the community structure and making full use of multifaceted information help learn better graph structure and more robust GCN parameters. Note that Geom-GCN does not perform well in most cases. One feasible reason is that it may fit the supervised settings in its original papers, where more supervised information is injected into parameter learning, and may not adapt well to the semi-supervised setting.

Table 3.3 Quantitative results ($\%\pm\sigma$) on node classification

Datasets	L/C	GCN	ChebNet	GAT	SAGE	LDS	ProGNN	GEN
Cora	20	81.7±0.8	81.9±0.4	82.3±1.0	80.1±0.5	82.5±1.2	80.9±0.9	**83.6**±0.4
	10	74.6±0.7	72.5±1.0	76.9±0.9	72.9±1.1	77.1±3.1	76.9±0.8	**77.8**±0.7
	5	71.0±0.7	66.6±2.3	75.0±0.7	68.4±1.7	75.7±2.9	75.1±0.5	**76.2**±1.3
Citeseer	20	70.9±0.6	70.0±0.9	72.0±0.9	71.8±0.7	72.3±1.1	68.8±0.8	**73.8**±0.6
	10	66.6±1.0	67.3±1.1	68.4±1.4	68.0±1.0	70.4±1.6	69.1±0.6	**72.4**±0.5
	5	53.5±0.8	51.7±2.3	61.8±1.9	55.4±1.0	68.1±0.5	56.6±1.5	**70.4**±2.7
Pubmed	20	79.4±0.4	78.2±1.0	77.9±0.6	73.6±2.2	78.2±1.8	78.0±0.8	**80.9**±0.9
	10	73.7±0.4	71.5±0.8	71.1±1.4	70.6±1.4	74.4±1.5	72.7±0.6	**75.6**±1.1
	5	73.0±1.4	69.4±1.4	70.2±0.7	70.2±1.2	72.8±1.3	70.6±1.7	**74.9**±2.0
Chame.	20	49.1±1.1	37.0±0.5	46.4±1.4	43.7±2.0	49.4±1.1	50.3±0.6	**50.4**±0.9
	10	44.2±0.7	32.5±0.8	45.0±2.0	41.7±1.9	44.9±1.3	45.5±1.2	**45.6**±1.1
	5	39.5±0.7	33.2±0.8	39.9±1.8	35.9±0.8	40.5±1.5	41.0±1.8	**41.4**±2.3
Squirrel	20	35.0±0.6	21.2±2.0	27.2±2.9	28.3±2.0	30.1±0.4	33.4±2.4	**35.5**±1.1
	10	33.0±0.4	18.8±1.2	27.1±1.2	25.9±2.9	29.4±0.9	32.9±0.4	**33.4**±1.1
	5	31.3±1.3	18.1±0.7	24.1±2.5	24.9±2.9	27.1±1.4	28.2±1.9	**32.7**±2.7
Actor	20	21.7±1.6	26.7±1.1	23.8±3.6	28.9±1.1	27.0±1.4	21.5±1.7	**35.3**±0.6
	10	20.8±1.0	22.3±1.1	22.7±3.6	22.2±2.5	25.7±1.3	22.2±0.7	**31.3**±2.2
	5	21.8±2.0	21.4±1.0	21.4±2.4	23.1±3.6	23.8±0.8	20.9±0.5	**30.5**±2.7

3.5 Interpreting and Unifying GNNs with An Optimization Framework

3.5.1 Overview

The well-designed propagation, i.e., message passing mechanism which has been demonstrated effective is the most fundamental part of GNNs. Although there are various propagation mechanisms, they basically utilize network topology and node features through aggregating node features along network topology. In view of this, one question naturally arises: albeit with different propagation strategies, is there a unified mathematical guideline that essentially governs the propagation mechanisms of different GNNs? If so, what is it? A well-informed answer to this question can provide a macroscopic view on surveying the relationships and differences between different GNNs in a principled way.

As the first contribution, we analyze the propagation process of several representative GNNs (e.g., GCN [102] and PPNP [103]), and abstract their commonalities. Surprisingly, we discover that they can be fundamentally summarized to a unified optimization framework with flexible graph convolutional kernels. The learned representation after propagation can be viewed as the optimal solution of the corresponding optimization objective implicitly. This unified framework consists of two terms: feature fitting term and graph Laplacian regularization term. The feature fitting term, building the relationship between node representation and original node features, is usually designed to meet different needs of specific GNNs. Graph Laplacian regularization term, playing the role of feature smoothing with topology, is shared by all these GNNs. For example, the propagation of GCN can be interpreted only by the graph Laplacian regularization term while PPNP needs another fitting term to constrain the similarity of the node representation and the original features.

Thanks to the macroscopic view on different GNNs provided by the proposed unified framework, the weakness of current GNNs is easy to be identified. As a consequence, the unified framework opens up new opportunities for designing novel GNNs. Traditionally, when we propose a new GNN model, we usually focus on designing a specific spectral graph filter or aggregation strategy. Now, the unified framework provides another new path to achieve this, i.e., the new GNN can be derived by optimizing an objective function. In this way, we clearly know the optimization objective behind the propagation process, making the new GNN more interpretable and more reliable. Here, with the proposed framework, we discover that existing works usually utilize naïve graph convolutional kernels for feature fitting functions, and then develop two novel flexible objective functions with adjustable kernels showing low-pass and high-pass filtering capabilities.

3.5.2 Preliminary

We consider graph convolutional operations on a graph $G = (V, \mathcal{E})$ with node set V and edge set $\mathcal{E}$, $n = |V|$ is the number of nodes. The nodes are described by the feature matrix $\mathbf{X} \in \mathbb{R}^{n \times f}$, where f is the dimension of node feature. Graph structure of G can be described by the adjacency matrix $\mathbf{A} \in \mathbb{R}^{n \times n}$ where $\mathbf{A}_{i,j} = 1$ if there is an edge between nodes i and j, otherwise 0. The diagonal degree matrix is denoted as $\mathbf{D} = diag(d_1, \ldots, d_n)$, where $d_i = \sum_j \mathbf{A}_{i,j}$. We use $\tilde{\mathbf{A}} = \mathbf{A} + \mathbf{I}$ to represent the adjacency matrix with added self-loop and $\tilde{\mathbf{D}} = \mathbf{D} + \mathbf{I}$. Then the normalized adjacency matrix is $\hat{\tilde{\mathbf{A}}} = \tilde{\mathbf{D}}^{-1/2} \tilde{\mathbf{A}} \tilde{\mathbf{D}}^{-1/2}$. Correspondingly, $\tilde{\mathbf{L}} = \mathbf{I} - \hat{\tilde{\mathbf{A}}}$ is the normalized symmetric positive semi-definite graph Laplacian matrix.

The Unified Framework. We summarize the K-layer propagation mechanisms mainly as the following two forms. For GNNs with layer-wise feature transformation (e.g., GCN [102]), the K-layer propagation process can be represented as

$$\mathbf{Z} = \mathbf{PROPAGATE}(\mathbf{X}; G; K) = \left\langle \mathbf{Trans}\left(\mathbf{Agg}\{G; \mathbf{Z}^{(k-1)}\}\right)\right\rangle_K, \qquad (3.44)$$

with $\mathbf{Z}^{(0)} = \mathbf{X}$ and $\mathbf{Z}$ is the output representation after the K-layer propagation. And $\langle\rangle_K$, usually depending on specific GNN models, represents the generalized combination operation after K convolutions. $\mathbf{Agg}\{G; \mathbf{Z}^{(k-1)}\}$ means aggregating the $(k-1)$-layer output $\mathbf{Z}^{(k-1)}$

along graph $\mathcal{G}$ for the kth convolutional operation, and $\textbf{\textit{Trans}}(\cdot)$ is the corresponding layer-wise feature transformation operation including non-linear activation function $ReLU()$ and layer-specific learnable weight matrix $\textbf{W}$.

Some deep graph neural networks (e.g., APPNP [103], DAGNN [119]) decouple the layer-wise $\textbf{\textit{Trans}}(\cdot)$ and $\textbf{\textit{Agg}}\{\mathcal{G}; \textbf{Z}^{(k-1)}\}$, and use a separated feature transformation before the consecutive aggregation steps:

$$\textbf{Z} = \textbf{PROPAGATE}(\textbf{X}; \mathcal{G}; K) = \left\langle \textbf{\textit{Agg}}\{\mathcal{G}; \textbf{Z}^{(k-1)}\}\right\rangle_K, \tag{3.45}$$

with $\textbf{Z}^{(0)} = \textbf{\textit{Trans}}(\textbf{X})$ and $\textbf{\textit{Trans}}(\cdot)$ can be any linear or non-linear transformation operation on original feature matrix $\textbf{X}$.

In addition, the combination operation $\langle\rangle_K$ is generally two-fold: for GNNs like GCN, SGC, and APPNP, $\langle\rangle_K$ directly utilizes the Kth layer output. And for GNNs using outputs from other layers, like JKNet and DAGNN, $\langle\rangle_K$ may represent pooling, concatenation, or attention operations on some (or all) outputs from K layers.

Various GNNs are proposed with different propagation mechanisms, in fact, they usually potentially aim at achieving two goals: encoding useful information from feature and utilizing the smoothing ability of topology, which can be formally formulated as the following optimization objective:

$$O = \min_{\textbf{Z}} \Big\{ \underbrace{\zeta \left\| \textbf{F}_1\textbf{Z} - \textbf{F}_2\textbf{H}\right\|_F^2}_{O_{fit}} + \underbrace{\xi tr(\textbf{Z}^T\tilde{\textbf{L}}\textbf{Z})}_{O_{reg}} \Big\}. \tag{3.46}$$

Here, ξ is a non-negative coefficient, ζ is usually chosen from $[0, 1]$, and $\textbf{H}$ is the transformation on original input feature matrix $\textbf{X}$. $\textbf{F}_1$ and $\textbf{F}_2$ are defined as arbitrary graph convolutional kernels. $\textbf{Z}$ is the propagated representation and corresponds to the final propagation result when minimizing the objective O.

In this unified framework, the first part O_{fit} is a fitting term which flexibly encodes the information in $\textbf{H}$ to the learned representation $\textbf{Z}$ through designing different graph convolutional kernels $\textbf{F}_1$ and $\textbf{F}_2$. Graph convolutional kernels $\textbf{F}_1$ and $\textbf{F}_2$ can be chosen from the $\textbf{I}$, $\hat{\textbf{A}}$, and $\tilde{\textbf{L}}$, showing the all-pass, low-pass, and high-pass filtering capabilities, respectively. The second term O_{reg} is a graph Laplacian regularization term constraining the learned representations of two connected nodes become similar, so that the homophily property can be captured, and O_{reg} comes from the following graph Laplacian regularization:

$$O_{reg} = \frac{\xi}{2} \sum_{i,j}^{n} \hat{\textbf{A}}_{i,j} \left\| \textbf{Z}_i - \textbf{Z}_j \right\|^2 = \xi tr(\textbf{Z}^T\tilde{\textbf{L}}\textbf{Z}). \tag{3.47}$$

The proposed framework shows a big picture of GNNs by mathematically modeling the objective optimization function. Considering that different existing GNNs can be fit into this framework, novel variations of GNNs can also be easily come up. All we need is to design the variables within this framework (e.g., different graph convolutional kernels $\mathbf{F}_1$ and $\mathbf{F}_2$) based on the specific scenarios, the corresponding propagation can be easily derived, and new GNNs architecture can be naturally designed. With one targeted objective function, the newly designed model is more interpretable and more reliable.

In [205], we theoretically prove that the propagation mechanisms of some typical GNNs are actually the special cases of our proposed unified framework, which enables us to interpret the current GNNs in a global perspective.

3.5.3   The GNN-LF/HF Method

Based on the unified framework, we find that most of the current GNNs simply set $\mathbf{F}_1$ and $\mathbf{F}_2$ as $\mathbf{I}$ in feature fitting term, implying that they require all original information in $\mathbf{H}$ to be encoded into $\mathbf{Z}$. However, in fact, the $\mathbf{H}$ may inevitably contain noise or uncertain information. We notice that JKNet has the propagation objective with $\mathbf{F}_2$ as $\hat{\mathbf{A}}$, which can encode the low-frequency information in $\mathbf{H}$ to $\mathbf{Z}$. While, in reality, the situation is more complex because it is hard to determine what information should be encoded, only considering one type of information cannot satisfy the needs of different downstream tasks, and sometimes high frequency or all information is even also helpful. In this section, we focus on designing novel $\mathbf{F}_1$ and $\mathbf{F}_2$ to flexibly encode more comprehensive information under the framework.

3.5.3.1   GNN with Low-Pass Filtering Kernel
We first consider building the relationship of $\mathbf{H}$ and $\mathbf{Z}$ in both original and low-pass filtering spaces.

Theorem 3.1 *With $\mathbf{F}_1 = \mathbf{F}_2 = \{\mu\mathbf{I} + (1 - \mu)\hat{\mathbf{A}}\}^{1/2}$, $\mu \in [1/2, 1)$, $\zeta = 1$ and $\xi = 1/\alpha - 1$, $\alpha \in (0, 2/3)$ in Eq. (3.46), the propagation process considering flexible low-pass filtering kernel on feature is*

$$O = min\{\left\|\{\mu\mathbf{I} + (1 - \mu)\hat{\mathbf{A}}\}^{1/2}(\mathbf{Z} - \mathbf{H})\right\|_F^2 + \xi tr(\mathbf{Z}^T \tilde{\mathbf{L}}\mathbf{Z})\}. \tag{3.48}$$

Note that μ is a balance coefficient, and we set $\mu \in [1/2, 1)$ so that $\mu\mathbf{I} + (1 - \mu)\hat{\mathbf{A}} = \mathbf{V}\Lambda\mathbf{V}^T$ is a symmetric and positive semi-definite matrix. Therefore, the matrix $\{\mu\mathbf{I} + (1 - \mu)\hat{\mathbf{A}}\}^{1/2} = \mathbf{V}\Lambda^{1/2}\mathbf{V}^T$ has a filtering behavior similar to that of $\mu\mathbf{I} + (1 - \mu)\hat{\mathbf{A}}$ in spectral domain. By adjusting the balance coefficient μ, the designed objective can flexibly constrain the similarity of $\mathbf{Z}$ and $\mathbf{H}$ in both original and low-pass filtering spaces, which is beneficial to meet the needs of different tasks.

Closed Solution. To minimize the objective function in Eq. (3.48), we set derivative of Eq. (3.48) with respect to $\mathbf{Z}$ to zero and derive the corresponding closed-form solution as follows:

$$\mathbf{Z} = \{\mu\mathbf{I} + (1 - \mu)\hat{\tilde{\mathbf{A}}} + (1/\alpha - 1)\tilde{\mathbf{L}}\}^{-1}\{\mu\mathbf{I} + (1 - \mu)\hat{\tilde{\mathbf{A}}}\}\mathbf{H}. \tag{3.49}$$

We can rewrite Eq. (3.49) using $\hat{\tilde{\mathbf{A}}}$ as

$$\mathbf{Z} = \{\{\mu + 1/\alpha - 1\}\mathbf{I} + \{2 - \mu - 1/\alpha\}\hat{\tilde{\mathbf{A}}}\}^{-1}\{\mu\mathbf{I} + (1 - \mu)\hat{\tilde{\mathbf{A}}}\}\mathbf{H}. \tag{3.50}$$

Iterative Approximation. Considering that the closed-form solution is computationally inefficient because of the matrix inversion, we can use the following iterative approximation solution instead without constructing the dense inverse matrix:

$$\mathbf{Z}^{(k)} = \frac{1 + \alpha\mu - 2\alpha}{1 + \alpha\mu - \alpha}\hat{\tilde{\mathbf{A}}}\mathbf{Z}^{(k-1)} + \frac{\alpha\mu}{1 + \alpha\mu - \alpha}\mathbf{H} + \frac{\alpha - \alpha\mu}{1 + \alpha\mu - \alpha}\hat{\tilde{\mathbf{A}}}\mathbf{H}, \tag{3.51}$$

which converge to the closed-form solution in Eq. (3.50) when $k \to \infty$, and with $\alpha \in (0, 2/3)$, all the coefficients are always positive.

3.5.3.2 Low-Pass Model Design

With the derived two propagation strategies in Eqs. (3.50) and (3.51), we propose two new GNNs in both closed and iterative forms. Note that we represent the proposed models as GNN with low-pass filtering graph convolutional kernel (**GNN-LF**).

GNN-LF-closed. Using the closed-form propagation matrix in Eq. (3.50), we define the following propagation mechanism with $\mu \in [1/2, 1)$, $\alpha \in (0, 2/3)$:

$$\begin{aligned}
\mathbf{Z} &= \mathbf{PROPAGATE}(\mathbf{X}; \mathcal{G}; \infty)_{LF-closed} \\
&= \{\{\mu + 1/\alpha - 1\}\mathbf{I} + \{2 - \mu - 1/\alpha\}\hat{\tilde{\mathbf{A}}}\}^{-1}\{\mu\mathbf{I} + (1 - \mu)\hat{\tilde{\mathbf{A}}}\}\mathbf{H}.
\end{aligned} \tag{3.52}$$

Here we first get a non-linear transformation result $\mathbf{H}$ on feature $\mathbf{X}$ with an MLP network $f_\theta(\cdot)$, and use the designed propagation matrix $\{\{\mu + 1/\alpha - 1\}\mathbf{I} + \{2 - \mu - 1/\alpha\}\hat{\tilde{\mathbf{A}}}\}^{-1}$ to propagate both $\mathbf{H}$ and $\mathbf{AH}$, then we can get the representation encoding feature information from both original and low-frequency spaces.

GNN-LF-iter. Using the iter-form propagation mechanism, we can design a deep and computationally efficient graph neural network with $\mu \in [1/2, 1)$, $\alpha \in (0, 2/3)$:

$$\begin{aligned}
\mathbf{Z} &= \mathbf{PROPAGATE}(\mathbf{X}; \mathcal{G}; K)_{LF-iter} \\
&= \left\langle \frac{1 + \alpha\mu - 2\alpha}{1 + \alpha\mu - \alpha}\hat{\tilde{\mathbf{A}}}\mathbf{Z}^{(k-1)} + \frac{\alpha\mu}{1 + \alpha\mu - \alpha}\mathbf{H} + \frac{\alpha - \alpha\mu}{1 + \alpha\mu - \alpha}\hat{\tilde{\mathbf{A}}}\mathbf{H} \right\rangle_K, \\
\mathbf{Z}^{(0)} &= \frac{\mu}{1 + \alpha\mu - \alpha}\mathbf{H} + \frac{1 - \mu}{1 + \alpha\mu - \alpha}\hat{\tilde{\mathbf{A}}}\mathbf{H}, \quad and \quad \mathbf{H} = f_\theta(\mathbf{X}).
\end{aligned} \tag{3.53}$$

We directly use the K-layer output as the propagation results. This iterative propagation mechanism can be viewed as layer-wise $\hat{\tilde{\mathbf{A}}}$-based neighborhood aggregation, with residual connection on feature matrix $\mathbf{H}$ and filtered feature matrix $\hat{\tilde{\mathbf{A}}}\mathbf{H}$. Note that we decouple the layer-wise transformation and aggregation process like [103, 119], which is beneficial to alleviate the over-smoothing problem.

3.5.3.3 GNN with High-Pass Filtering Kernel

Similar with GNN-LF, we now consider preserving the similarity of $\mathbf{H}$ and $\mathbf{Z}$ in both original and high-pass filtering spaces. For neatness of the subsequent analysis, we choose the following objective:

Theorem 3.2 *With* $F_1 = F_2 = \{I + \beta\tilde{L}\}^{1/2}$, $\beta \in (0, \infty)$, $\zeta = 1$ *and* $\xi = 1/\alpha - 1$, $\alpha \in (0, 1]$ *in Eq. (3.46), the propagation process considering flexible high-pass convolutional kernel on feature is*

$$O = \min_{\mathbf{Z}} \left\{ \left\| \{I + \beta\tilde{L}\}^{1/2}(\mathbf{Z} - H) \right\|_F^2 + \xi tr(\mathbf{Z}^T \tilde{L}\mathbf{Z}) \right\}. \tag{3.54}$$

Analogously, β is also a balance coefficient, and we set $\beta \in (0, \infty)$ so that $\mathbf{I} + \beta\tilde{\mathbf{L}} = V^*\Lambda^*V^{*T}$ is also a symmetric and positive semi-definite matrix and the matrix $\{\mathbf{I} + \beta\tilde{\mathbf{L}}\}^{1/2} = V^*\Lambda^{*1/2}V^{*T}$ has a filtering behavior similar to that of $\{\mathbf{I} + \beta\tilde{\mathbf{L}}\}$. As can be seen in Eq. (3.54), by adjusting the balance coefficient β, the designed objectives can flexibly constrain the similarity of $\mathbf{Z}$ and $\mathbf{H}$ in both original and high-frequency spaces.

Closed Solution. We calculate the closed-form solution as

$$\mathbf{Z} = \left\{ \mathbf{I} + (\beta + 1/\alpha - 1)\tilde{\mathbf{L}} \right\}^{-1} \{\mathbf{I} + \beta\tilde{\mathbf{L}}\}\mathbf{H}, \tag{3.55}$$

it also can be rewritten as

$$\mathbf{Z} = \left\{ (\beta + 1/\alpha)\mathbf{I} + (1 - \beta - 1/\alpha)\hat{\tilde{\mathbf{A}}} \right\}^{-1} \{\mathbf{I} + \beta\tilde{\mathbf{L}}\}\mathbf{H}. \tag{3.56}$$

Iterative Approximation. Considering it is inefficient to calculate the inverse matrix, we give the following iterative approximation solution without constructing the dense inverse matrix:

$$\mathbf{Z}^{(k)} = \frac{\alpha\beta - \alpha + 1}{\alpha\beta + 1}\hat{\tilde{\mathbf{A}}}\mathbf{Z}^{(k-1)} + \frac{\alpha}{\alpha\beta + 1}\mathbf{H} + \frac{\alpha\beta}{\alpha\beta + 1}\tilde{\mathbf{L}}\mathbf{H}. \tag{3.57}$$

3.5.3.4 High-Pass Model Design

With the derived two propagation strategies in Eqs. (3.56) and (3.57), we propose two new GNNs in both closed and iterative forms. Similarly, we use **GNN-HF** to denote GNN with high-pass filtering graph convolutional kernels.

GNN-HF-closed. Using the closed-form propagation matrix in Eq. (3.56), we define the closed-form propagation mechanism:

$$Z = \textbf{PROPAGATE}(X; \mathcal{G}; \infty)_{HF-closed}$$

$$= \left\{ (\beta + 1/\alpha)I + (1 - \beta - 1/\alpha)\hat{\tilde{A}} \right\}^{-1} \{I + \beta \tilde{L}\} H. \tag{3.58}$$

Note that $\beta \in (0, \infty)$ and $\alpha \in (0, 1]$. By applying the propagation matrix $\{(\beta + 1/\alpha)I + (1 - \beta - 1/\alpha)\hat{\tilde{A}}\}^{-1}$ directly on both H and $\tilde{L}H$ matrix, then we can get the representation encoding feature information from both original and high-frequency spaces.

GNN-HF-iter. Using the iterative propagation mechanism, we have a deep and computationally efficient graph neural networks with $\beta \in (0, \infty)$ and $\alpha \in (0, 1]$.

$$Z = \textbf{PROPAGATE}(X; \mathcal{G}; K)_{HF-iter}$$

$$= \left\langle \frac{\alpha\beta - \alpha + 1}{\alpha\beta + 1} \hat{\tilde{A}} Z^{(k-1)} + \frac{\alpha}{\alpha\beta + 1} H + \frac{\alpha\beta}{\alpha\beta + 1} \tilde{L}H \right\rangle_K, \tag{3.59}$$

$$Z^{(0)} = \frac{1}{\alpha\beta + 1} H + \frac{\beta}{\alpha\beta + 1} \tilde{L}H, \quad and \quad H = f_\theta(X).$$

We directly use the K-layer output as the propagation results. Similarly, this iterative propagation mechanism can be viewed as layer-wise $\hat{\tilde{A}}$-based neighborhood aggregation, and residual connection on both feature matrix H and high-frequency filtered feature matrix $\tilde{L}H$. And we also decouple the layer-wise transformation and aggregation process during propagation.

3.5.4 Experiments

3.5.4.1 Experimental Settings

To evaluate the effectiveness of our proposed GNN-LF/HF, we conduct experiments on six benchmark datasets. (1) Cora, Citeseer, Pubmed [102]: Three standard citation networks where nodes represent documents, edges are citation links, and features are the bag-of-words representation of the document. (2) ACM [200]: Nodes represent papers and there is an edge if two papers have same authors. Features are the bag-of-words representations of paper keywords. The three classes are *Database, Wireless Communication, DataMining*. (3) Wiki-CS [131]: A dataset derived from Wikipedia, in which nodes represent CS articles, edges are hyperlinks, and different classes mean different branches of the files. (4) MS Academic [103]: A co-authorship Microsoft Academic Graph, where nodes represent authors, edges are co-authorships, and node features represent keywords from authors' papers.

3.5.4.2 Node Classification Results

We evaluate the effectiveness of GNN-LF/HF against several state-of-the-art baselines on semi-supervised node classification task. We use accuracy (ACC) metric for evaluation, and

Table 3.4 Node classification results (%)

Model	Dataset					
	Cora	Citeseer	Pubmed	ACM	Wiki-CS	MS academic
MLP	57.79±0.11	61.20±0.08	73.23±0.05	77.39±0.11	65.66±0.20	87.79±0.42
LP	71.50±0.00	50.80±0.00	72.70±0.00	63.30±0.00	34.90±0.00	74.10±0.00
ChebNet	79.92±0.18	70.90±0.37	76.98±0.16	79.53±1.24	63.24±1.43	90.76±0.73
GAT	82.48±0.31	72.08±0.41	79.08±0.22	88.24±0.38	74.27±0.63	91.58±0.25
GraphSAGE	82.14±0.25	71.80±0.36	79.20±0.27	87.57±0.65	73.17±0.41	91.53±0.15
IncepGCN	81.94±0.94	69.66±0.29	78.88±0.35	87.75±0.61	60.54±1.06	75.45±0.49
GCN	82.41±0.25	70.72±0.36	79.40±0.15	88.38±0.51	71.97±0.51	92.17±0.11
SGC	81.90±0.23	72.21±0.22	78.30±0.14	87.56±0.34	72.43±0.28	88.35±0.36
PPNP	83.34±0.20	71.73±0.30	80.06±0.20	89.12±0.17	74.53±0.36	92.27±0.23
APPNP	83.32±0.42	71.67±0.48	80.05±0.27	89.04±0.21	74.30±0.50	92.25±0.18
JKNet	81.19±0.49	70.69±0.88	78.60±0.25	88.11±0.36	60.90±0.92	87.26±0.23
GNN-LF-closed	83.70±0.14	71.98±0.33	80.34±0.18	89.43±0.20	**75.50±0.56**	**92.79±0.15**
GNN-LF-iter	83.53±0.24	71.92±0.24	80.33±0.20	89.37±0.40	75.35±0.24	92.69±0.20
GNN-HF-closed	**83.96±0.22**	**72.30±0.28**	80.41±0.25	89.46±0.30	74.92±0.45	92.47±0.23
GNN-HF-iter	83.79±0.29	72.03±0.36	**80.54±0.25**	**89.59±0.31**	74.90±0.37	92.51±0.16

report the average ACC with uncertainties showing the 95% confidence level calculated by bootstrapping in Table 3.4. We have the following observations:

(1) GNN-LF and GNN-HF consistently outperform all the state-of-the-art baselines on all datasets. The best and the runner-up results are always achieved by GNN-LF/HF, which demonstrates the effectiveness of our proposed model. From the perspective of the unified objective framework, it is easy to check that GNN-LF/HF not only keep the representation same with the original features, but also consider capturing their similarities based on low-frequency or high-frequency information. These two relations are balanced so as to extract more meaningful signals and thus perform better.

(2) From the results of the closed and iterative versions of GNN-LF/HF, we can see that using 10 propagation depth for GNN-LF-iter/GNN-HF-iter is able to effectively approximate the GNN-LF-closed/GNN-HF-closed. As for performance comparisons between GNN-LF and GNN-HF, we find that it is hard to determine which is the best, since which filter works better may depend on the characteristic of different datasets. But in summary, flexibly and

comprehensively considering multiple information in a GNN model can always achieve
satisfactory results on different networks.

(3) In addition, PPNP/APPNP always perform better than GCN/SGC since their objective also considers a fitting term to help find important information from features during
propagation. On the other hand, APPNP outperforms JKNet mainly because that its propagation process takes full advantage of the original features and APPNP even decouples the
layer-wise non-linear transformation operations [119] without suffering from performance
degradation. Actually, the above differences of models and explanations for results can be
easily drawn from our unified framework.

3.6 Conclusion

In this chapter, we introduce homogeneous graph neural networks, the most fundamental
part of which is the design of message-passing functions. Specifically, we introduce three
representative methods: AM-GCN, which enhances GCN's ability to fuse attribute and
topology information; FAGCN, which helps existing GNNs get rid of the low-passing filtering limitation; and GEN, which learns a better message-passing topology for GNNs. With
sufficient experiments, the effectiveness of the proposed methods is well proved. Further,
we show that the message-passing functions of most existing GNNs can be summarized into
a unified closed-form framework with a flexible iterative algorithm. Based on the framework, we design two novel GNNs: GNN-LF and GNN-HF, which can perform low-pass
and high-pass filtering, respectively. We believe that these discoveries can help researchers
understand principles behind message passing.

3.7 Further Reading

The research of homogeneous GNNs is the most fundamental part of GNNs. For a more
comprehensive introduction of homogeneous GNNs, we suggest the audiences read the
recent surveys on GNNs [210, 245]. Generally, we can divide existing homogeneous GNNs
into two categories: spectral and spatial. These two methods give us different ways of
designing the message-passing functions.

Spectral methods treat the graph topology as a filter and design new topology through
spectral graph theory. After passing the messages on the graph filter, the essential frequency
information in node features is preserved and the noise is filtered. Sections 3.3 and 3.5 belong
to this category. The first spectral GNN is proposed by Yann LeCun [12], suffering from
the high computation cost in eigendecomposition. ChebyNet [37] solves this problem by
utilizing the character that symmetric matrix is diagonalizable. And GCN [102] simplifies
ChebNet through a first-order approximation.

   Spatial methods focus on designing powerful message-passing functions to aggregate the information of neighbors. Sections 3.2 and 3.4 belong to this category. In particular, spatial methods will draw lessons from the state-of-the-art neural network architectures. For example, GAT [189] uses the attention mechanism to learn the importance of different neighbors, PPNP [103] uses the skip-connections to deal with the over-smoothing problem in GNNs, and AdaGCN [176] incorporates a traditional boosting method into GNNs.

Heterogeneous Graph Neural Networks

4

Ruijia Wang

4.1 Introduction

Previous GNNs mainly focus on the homogeneous graph, while heterogeneous graph (HG) is ubiquitous ranging from bibliographic and social networks to transportation and telecommunication systems [167]. With heterogeneous types of nodes and relations, HGs are able to model complex interactions and immensely rich semantics in real-world scenarios.

Recently, heterogeneous graph neural networks (HGNNs) [84, 195, 200, 203, 233] have achieved great success in dealing with HGs, because they effectively combine the mechanism of message passing with complex heterogeneity. These HGNNs often boil down to a two-step aggregating framework in a hierarchical manner: (a) aggregating neighbors extracted by single meta-path in node-level; (b) aggregating rich semantics via multiple meta-paths in semantic level. Until now, HGNNs have significantly promoted the development of HG analysis toward real-world applications, e.g., recommender [45] and security system [47].

In this chapter, we introduce three representative HGNNs dedicated to two key issues, i.e., deep degradation phenomenon and discriminative power. Specifically, Heterogeneous graph Propagation Network (HPN) [90] theoretically analyzes the deep degradation phenomenon in HGNNs and proposes a novel convolution layer to alleviate this semantic confusion. In terms of the discriminative power, distance encoding-based heterogeneous graph neural network (DHN) [89] injects heterogeneous distance encoding into aggregation, while self-supervised heterogeneous graph neural network with co-contrastive learning (HeCo) [201] employs a cross-view contrastive mechanism to capture both of local and high-order structures simultaneously.

R. Wang (✉)
Beijing University of Posts and Telecommunications, Beijing, China

4.2   Heterogeneous Graph Propagation Network

4.2.1   Overview

When applying HGNNs in practice, an important phenomenon is discovered, called *semantic confusion*. Similar to over-smoothing in homogeneous GNNs [116, 216], semantic confusion means HGNNs inject confused semantics extracted via multiple meta-paths into node embedding. Figure 4.1a shows the clustering performance of HAN on ACM academic graph [200]. It displays that with the growth of model depth, the performance of HAN is getting worse and worse. The reasons for semantic confusion are two folds: first, with the development of model depth, different nodes will connect to the same meta-path-based neighbors. Second, multiple meta-path combinations in semantic-level aggregating fuse multiple indistinguishable semantics. Semantic confusion makes HGNNs hard to become a really deep model, which severely limits their representation capabilities and hurts the performance of downstream tasks.

Therefore, heterogeneous graph propagation network (HPN) theoretically analyzes the semantic confusion in HGNNs and proves that HGNNs and multiple meta-paths-based random walk [110] are essentially equivalent. Then HPN alleviates semantic confusion from the perspective of multiple meta-paths-based random walk. Specifically, HPN contains two parts: semantic propagation mechanism and semantic fusion mechanism. Besides aggregating information from meta-path-based neighbors, the semantic propagation mechanism also absorbs the node's local semantics with a proper weight. So even with more hidden

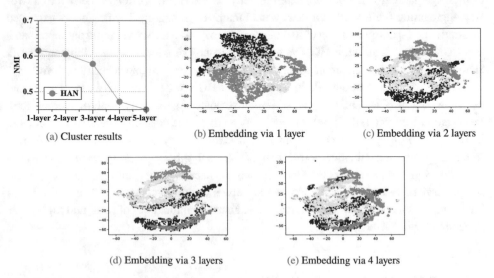

(a) Cluster results   (b) Embedding via 1 layer   (c) Embedding via 2 layers

(d) Embedding via 3 layers   (e) Embedding via 4 layers

Fig. 4.1 The clustering results and visualization of paper embeddings via HAN with different layers. Each point denotes one paper and corresponding color indicates the label (i.e., research areas)

layers, semantic propagation mechanism can capture the characteristics of each node. And the semantic fusion mechanism learns the importance of meta-paths and fuses them for comprehensive node embedding.

4.2.2   The HPN Method

4.2.2.1 Relationship Between HGNNs and Meta-Paths-Based Random Walk

As a classical heterogeneous graph algorithm, multiple meta-paths-based random walk [110] mainly contains single meta-path-based random walk and multiple meta-path combinations.Given a meta-path Φ, the element $\mathbf{M}_{ij}^{\Phi}$ of meta-path-based probability matrix $\mathbf{M}^{\Phi}$ denotes the transition probability from node i to j via meta-path Φ. Then, the k-step single meta-path-based random walk is defined as

$$\pi^{\Phi,k} = \mathbf{M}^{\Phi} \cdot \pi^{\Phi,k-1}, \tag{4.1}$$

where $\pi^{\Phi,k}$ denotes the distribution of k-step single meta-path-based random walk. Considering meta-path set $\{\Phi_1, \Phi_2, \ldots, \Phi_P\}$ and their weights $\{w_{\Phi_1}, w_{\Phi_2}, \ldots, w_{\Phi_P}\}$, the k-step multiple meta-paths-based random walk is defined as

$$\pi^k = \sum_{p=1}^{P} w_{\Phi_p} \cdot \pi^{\Phi_p,k}, \tag{4.2}$$

where π^k denotes the distribution of k-step multiple meta-paths-based random walk. For k-step single meta-path-based random walk:

Theorem 4.1 *Assuming a heterogeneous graph is aperiodic and irreducible, if taking the limit $k \to \infty$, then k-step meta-path-based random walk will converge to a meta-path specific limit distribution $\pi^{\Phi,\lim}$ which is independent of nodes:*

$$\pi^{\Phi,\lim} = \mathbf{M}^{\Phi} \cdot \pi^{\Phi,\lim}. \tag{4.3}$$

Since $\mathbf{M}^{\Phi}$ is a probability matrix, so the meta-path-based random walk is a Markov chain. The convergence of Markov chain shows that $\pi^{\Phi,k}$ will converge to a limit distribution $\pi^{\Phi,\lim}$ if taking the limit $k \to \infty$. Obviously, $\pi^{\Phi,\lim}$ only depends on $\mathbf{M}^{\Phi}$ and is independent of nodes.

Different nodes connected via some relationships will influence each other and [216] demonstrates the influence distribution between two nodes is proportional to random walk distribution, shown as the following theorem:

Theorem 4.2 ([216]) *For the aggregation models (e.g., graph neural networks) on homogeneous graph, if the graph is aperiodic and irreducible, then the influence distribution I_i of node i is equivalent, in expectation, to the k-step random walk distribution.* □

By Theorems 4.1 and 4.2, it is concluded that the influence distribution revealed by single meta-path-based random walk is independent of nodes. So node-level aggregation in HGNNs is essentially equivalent to meta-path-based random walk if activate function is a linear function. Based on the above analysis, if stacking infinite layers in node-level aggregating, the learned node embeddings $\mathbf{Z}^\Phi$ will only be influenced by the meta-path Φ and therefore are independent of nodes. So the learned node embeddings cannot capture the characteristics of each node and therefore are indistinguishable. For k-step multiple meta-paths-based random walk:

Theorem 4.3 *Assuming k-step single meta-path-based random walk is independent of each other, if taking the limit $k \to \infty$, then the limit distribution of k-step multiple meta-paths-based random walk is a weighted combination of single meta-path-based random walk limit distribution, shown as follows:*

$$\pi^{\lim} = \sum_{p=1}^{P} w_{\Phi_p} \cdot \pi^{\Phi_p, \lim}. \tag{4.4}$$

Proof Since k-step single meta-path-based random walk is independent of each other, according to the properties of limits including Sum Rule and Constant Multiple Rule [258]:

$$\begin{aligned}
\pi^{\lim} &= \lim_{k \to \infty} \sum_{p=1}^{P} w_{\Phi_p} \cdot \pi^{\Phi_p, k} = \sum_{p=1}^{P} w_{\Phi_p} \cdot \lim_{k \to \infty} \pi^{\Phi_p, k} \\
&= \sum_{p=1}^{P} w_{\Phi_p} \cdot \pi^{\Phi_p, \lim}.
\end{aligned} \tag{4.5}$$

It shows that the meta-path combination can only change the position of limit distribution, but convergence of limit distribution remains unchanged. □

By Theorems 4.2 and 4.3, it is concluded that the final node embeddings learned via both node- and semantic level only influenced by a set of meta-paths, thus remain indistinguishable. To alleviate this semantic confusion phenomenon, HPN improves the current HGNN architectures in node-level or semantic level.

4.2.2.2 The Proposed Model: HPN

The proposed HPN mainly consists of a semantic propagation mechanism and a semantic fusion mechanism. Inspired by meta-path-based random walk with restart, semantic
propagation mechanism emphasizes node's local semantics. Semantic fusion mechanism is
able to learn the importance of meta-paths and get the optimally weighted combination of
semantic-specific node embedding.

Semantic Propagation Mechanism. Given one meta-path Φ, the semantic propagation
mechanism $\mathcal{P}_\Phi$ first projects node into semantic space via semantic projection function f_Φ.
Then, it aggregates information from meta-path-based neighbors via semantic aggregation
function g_Φ, shown as follows:

$$\mathbf{Z}^\Phi = \mathcal{P}_\Phi(\mathbf{X}) = g_\Phi(f_\Phi(\mathbf{X})), \tag{4.6}$$

where $\mathbf{X}$ denotes initial feature matrix and $\mathbf{Z}^\Phi$ denotes semantic-specific node embedding. To
handle heterogeneity graph, the semantic projection function f_Φ projects node into semantic
space, shown as follows:

$$\mathbf{H}^\Phi = f_\Phi(\mathbf{X}) = \sigma(\mathbf{X} \cdot \mathbf{W}^\Phi + \mathbf{b}^\Phi), \tag{4.7}$$

where $\mathbf{H}^\Phi$ is the projected node feature matrix, $\mathbf{W}^\Phi$ and $\mathbf{b}^\Phi$ denote weight matrix and bias
vector for meta-path Φ, respectively. Note that $\mathbf{H}^\Phi$ can also be viewed as the 0-order node
embedding $\mathbf{Z}^{\Phi,0}$, revealing the characteristics of each node. To alleviate semantic confusion,
semantic aggregation function g_Φ is designed as follows:

$$\mathbf{Z}^{\Phi,k} = g_\Phi(\mathbf{Z}^{\Phi,k-1}) = (1 - \gamma) \cdot \mathbf{M}^\Phi \cdot \mathbf{Z}^{\Phi,k-1} + \gamma \cdot \mathbf{H}^\Phi, \tag{4.8}$$

where $\mathbf{Z}^{\Phi,k}$ denotes node embedding learned by kth layer semantic propagation mechanism,
and $\mathbf{M}^\Phi \cdot \mathbf{Z}^{\Phi,k-1}$ means aggregating information from meta-path-based neighbors. Here γ
is a weight scalar which indicates the importance of characteristic of node in aggregating
process.

Why semantic aggregation function g_Φ works. Here the relationship is established
between semantic aggregation function g_Φ and k-step meta-path-based random walk with
restart. The k-step meta-path-based random walk with restart for node i is defined as

$$\pi^{\Phi,k}(i) = (1 - \gamma) \cdot \mathbf{M}^\Phi \cdot \pi^{\Phi,k-1}(i) + \gamma \cdot i, \tag{4.9}$$

where i is a one-hot vector of node i, γ means the restart probability. For k-step meta-path-
based random walk with restart:

Theorem 4.4 *Assuming a heterogeneous graph is aperiodic and irreducible, if taking the
limit $k \to \infty$, then* k-*step meta-path-based random walk with restart will converge to*
$\pi^{\Phi,\lim}(i)$ *which is related to the start node i:*

$$\pi^{\Phi,\lim}(i) = \gamma \cdot (\mathbf{I} - (1 - \gamma) \cdot \mathbf{M}^{\Phi})^{-1} \cdot i. \tag{4.10}$$

Proof If taking the limit $k \to \infty$:

$$\pi^{\Phi,\lim}(i) = (1 - \gamma) \cdot \mathbf{M}^{\Phi} \cdot \pi^{\Phi,\lim}(i) + \gamma \cdot i. \tag{4.11}$$

Solving Eq. (4.11):

$$\pi^{\Phi,\lim}(i) = \gamma \cdot (\mathbf{I} - (1 - \gamma) \cdot \mathbf{M}^{\Phi})^{-1} \cdot i. \tag{4.12}$$

Obviously, $\pi^{\Phi,\lim}(i)$ is related to node i. $\square$

By Theorems 4.2 and 4.4, it is concluded that the influence distribution revealed by meta-path-based random walk with restart is related to nodes. The semantic aggregation function g_{Φ} absorbs node's local semantics and makes semantic-specific node embedding $\mathbf{Z}^{\Phi,k}$ distinguish from each other.

Semantic Fusion Mechanism. Given a set of meta-paths $\{\Phi_1, \Phi_2, \ldots, \Phi_P\}$, P group semantic-specific node embeddings $\{\mathbf{Z}^{\Phi_1}, \mathbf{Z}^{\Phi_2}, \ldots, \mathbf{Z}^{\Phi_P}\}$ are obtained. Then the semantic fusion mechanism $\mathcal{F}$ fuses them for the specific task. Taking P groups of semantic-specific node embeddings as input, the final node embedding $\mathbf{Z}$ learned via semantic fusion mechanism $\mathcal{F}$, shown as follows:

$$\mathbf{Z} = \mathcal{F}(\mathbf{Z}^{\Phi_1}, \mathbf{Z}^{\Phi_2}, \ldots, \mathbf{Z}^{\Phi_P}). \tag{4.13}$$

Intuitively, not all meta-paths should be treated equally. To learn the importance of meta-paths, each semantic-specific node embedding is projected into the same latent space and semantic fusion vector $\mathbf{q}$ is adopted to learn the importance of meta-paths. The importance of meta-path Φ_p, denoted as w_{Φ_p}, is defined as

$$w_{\Phi_p} = \frac{1}{|\mathcal{V}|} \sum_{i \in \mathcal{V}} \mathbf{q}^{\mathrm{T}} \cdot \tanh(\mathbf{W} \cdot \mathbf{z}_i^{\Phi_p} + \mathbf{b}), \tag{4.14}$$

where $\mathbf{W}$ and $\mathbf{b}$ denote weight matrix and bias vector, respectively. Note that all parameters in semantic fusion mechanism are shared for all nodes and semantics. After obtaining the importance of meta-paths, they are normalized via softmax function to get the weight of each meta-path β_{Φ_p}:

$$\beta_{\Phi_p} = \frac{\exp(w_{\Phi_p})}{\sum_{p=1}^{P} \exp(w_{\Phi_p})}. \tag{4.15}$$

With the learned weights as coefficients, P semantic-specific embeddings are fused to obtain the final embedding $\mathbf{Z}$ as follows:

$$\mathbf{Z} = \sum_{p=1}^{P} \beta_{\Phi_p} \cdot \mathbf{Z}^{\Phi_p}. \tag{4.16}$$

Loss Functions. For semi-supervised node classification, parameters is updated by cross-entropy loss in HPN:

$$\mathcal{L} = -\sum_{l \in \mathcal{Y}_L} \mathbf{Y}_l \cdot \ln(\mathbf{Z}_l \cdot \mathbf{C}), \tag{4.17}$$

where $\mathbf{C}$ is a projection matrix, $\mathcal{Y}_L$ is the set of labeled nodes, $\mathbf{Y}_l$ and $\mathbf{Z}_l$ are the label vector and embedding of the labeled node l, respectively.

For unsupervised node recommendation, BPR loss with negative sampling [77, 199] is leveraged to update parameters in HPN:

$$\mathcal{L} = -\sum_{(u,v) \in \Omega} \log \sigma \left(\mathbf{z}_u^\top \mathbf{z}_v \right) - \sum_{(u,v') \in \Omega^-} \log \sigma \left(-\mathbf{z}_u^\top \mathbf{z}_{v'} \right), \tag{4.18}$$

where $(u, v) \in \Omega$ and $(u, v') \in \Omega^-$ denote the set of observed (positive) node pairs and the set of negative node pairs sampled from all unobserved node pairs, respectively.

4.2.3   Experiments

4.2.3.1 Datasets and Baselines

Datasets. Experiments are conducted on real-world heterogeneous graphs *Yelp*,[1] *ACM*,[2] *IMDB*[3] and *MovieLens*.[4] The detailed descriptions are shown in Table 4.1.

Baselines. HPN is compared with some state-of-the-art baselines, including the heterogeneous network embedding (i.e., mp2vec [42] and HERec [166]), homogeneous GNNs (i.e., GCN [102] and PPNP [103]), and HGNNs (i.e., HAN [200] and MAGNN [56]). Meanwhile, two variants HPN_{pro} (restart probability $\gamma = 0$) and HPN_{fus} (simple average over all meta-paths) is set as baselines.

4.2.3.2 Node Clustering

In order to compare unsupervised models (i.e., metapath2vec and HERec) with semi-supervised models (i.e., GCN, PPNP, MAGNN, HAN and HPN), following the previous work [200], the learned node embeddings of all models are obtained via feedforward, then K-Means is applied to test their effectiveness. Metrics *NMI* and *ARI* evaluate the clustering task and the averaged results of 10 runs are reported in Table 4.2.

As can be seen, HPN performs significantly better than all baselines. It shows the importance of alleviating semantic confusion in HGNNs. It is also found HGNNs outperform homogeneous GNNs because they can capture rich semantics and describe the character-

---

[1] https://www.yelp.com.
[2] http://dl.acm.org/.
[3] https://www.kaggle.com/carolzhangdc/imdb-5000-movie-dataset.
[4] https://grouplens.org/datasets/movielens/.

Table 4.1 Statistics of the datasets

Dataset	A-B	#A	#B	#A-B	Fea.	Meta-paths/ Semantics
IMDB	Movie-Actor	4780	5841	14340	1232	Movie-Actor-Movie (*MAM*)
	Movie-Director	4780	2269	4780		Movie-Director-Movie (*MDM*)
ACM	Paper-Author	3025	5835	3025	1870	Paper-Author-Paper (*PAP*)
	Paper-Subject	3025	56	3025		Paper-Subject-Paper (*PSP*)
Yelp	Bus.-Category	4463	733	17123	144	Business-Category-Business (*BCB*)
	Bus.-Attribute	4463	144	82705		Business-Attribute-Business (*BAB*)
	Bus.-User	4463	29383	44816		Business-User-Business (*BUB*)
ML	User-Movie	943	1682	100000	18	User-Movie (*UM*)
	User-User	943	943	47150		User-User (*UU*)

Table 4.2 Quantitative results (%) on the node clustering task

Datasets	Metrics	mp2vec	HERec	GCN	PPNP	HAN	MG	HPN_{pro}	HPN_{fus}	HPN
Yelp	NMI	42.04	0.30	32.58	40.60	45.46	47.56	44.36	12.86	**48.90**
	ARI	38.27	0.41	23.30	37.72	41.39	43.24	42.57	10.54	**44.89**
ACM	NMI	21.22	40.70	51.40	61.68	61.56	64.12	65.60	67.55	**68.21**
	ARI	21.00	37.13	53.01	65.15	64.39	66.29	69.30	71.53	**72.33**
IMDB	NMI	1.20	1.20	5.45	10.20	10.87	11.79	9.45	12.01	**12.31**
	ARI	1.70	1.65	4.40	8.20	10.01	10.32	8.02	12.32	**12.55**

istic of node more comprehensively. Note that the performance of HPN_{pro} and HPN_{fus} both show different degradations, which imply the importance of semantic propagation mechanism and semantic fusion mechanism.

4.2.3.3 Robustness to Model Depth

A salient property of HPN is the incorporation of the semantic propagation mechanism. Comparing to the previous HGNNs, HPN can stack more layers and learn more representative node embedding. To show the superiority of semantic propagation in HPN, HAN and HPN with 1, 2, 3, 4, and 5 layers are tested, shown in Fig. 4.2. As can be seen, with the growth of model depth, the performance of HAN performs worse on both ACM and IMDB.

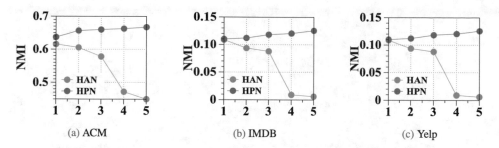

Fig. 4.2 Clustering results of HAN/HPN with 1, 2, 3, 4, and 5 layers

On the other hand, with the growth of model depth, the performance of HPN is getting better, indicating that the semantic propagation mechanism is able to effectively alleviate the semantic confusion. So even stacking for more layers, the node embeddings learned via HPN are still distinguishable.

The more detailed description and experiment validation can be seen in [90].

4.3   Heterogeneous Graph Neural Network with Distance Encoding

4.3.1   Overview

The prevailing instantiations of HGNNs focus on how to handle heterogeneity and learn *individual* node embedding via aggregating its structural neighbors. However, such a learning paradigm of HGNNs fails to establish the *correlation* among nodes [115]. As shown in Fig. 4.3b, both nodes p_2 and p_3 individually aggregate one author and two terms. Given two node pairs (p_1, p_2) and (p_1, p_3), HGNNs predict the same existence probability (i.e., $\hat{y}_{p_1,p_2} = \hat{y}_{p_1,p_3}$), shown in Fig. 4.3c. In fact, node p_1 is relatively closer to node p_2 than node p_3, so link (p_1, p_2) is more likely to exist. One important reason causing the above limited representational power of HGNNs lies in structural neighbor aggregating for each node individually, largely ignoring the correlation between nodes (e.g., distance) [115, 151, 226]. Recently, several works [115, 226] integrate diverse correlations into the learning process of homogeneous GNNs [115]. However, these correlation modeling technologies do not consider diverse types of edges, so they cannot be directly applied to HGs.

Establishing the correlation between nodes in the HGs encounters much more challenges because there exist diverse types of connections between nodes. Traditional measures (e.g., SPD) cannot sufficiently measure the correlation between nodes on HGs, because they only focus on the path length and largely ignore the influence of the path types. As shown in Fig. 4.3d, node pair (p_2, p_1) shows a different correlation from node pair (p_2, p_3), because they are connected via different types of paths, while the traditional SPD [115] assigns the same distance to them.

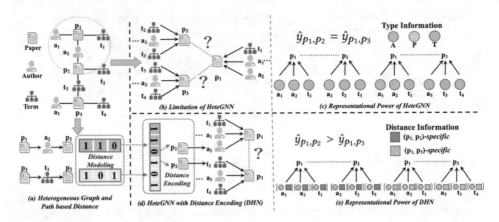

Fig. 4.3 An academic HG and the comparison between HGNNs and DHN

Therefore, heterogeneous distance encoding (HDE) is proposed to handle the above challenges. By injecting HDE into the neighbor aggregating process of HGNNs, a novel distance encoding-based heterogeneous graph neural network (DHN) is further proposed. Specifically, DHN first formulates the heterogeneous shortest path distance, and designs HDE to encode such distances among multiple nodes. After that, DHN injects the encoded correlation into the aggregating process to learn more expressive representations for link prediction.

4.3.2 The DHN Method

Definition (*Heterogeneous Enclosing Subgraph*) Given a target node set $S \subset \mathcal{V}$, the k-hop heterogeneous enclosing subgraph for S, denoted as $\mathcal{G}_S^k$, is the subgraph induced from the HG by the union of the k-hop neighbors of all nodes in S.   □

Example. Taking node pair (p_1, a_2) as an example, the 1-hop heterogeneous enclosing subgraph $\mathcal{G}_{\{p_1,a_2\}}^1$ is shown within the green circle in Fig. 4.3a, which consists of p_1's 1-hop neighbors $\{a_1, t_1, p_1, a_2\}$, a_2's 1-hop neighbors $\{p_1, p_2, a_2\}$, as well as the induced edges.

Definition (*Shortest Path Distance*) Given a node pair (u, v), a path ρ is defined as a node sequence from node v to u, denoted as $\rho_{v \leadsto u} = (w_0, w_1, \ldots, w_p)$, $w_0 = v$, $w_p = u$. Among all possible paths $\mathcal{P}_{v \leadsto u}$, the length of the shortest path (SPD), denoted as $spd(u|v) \in \mathbb{R}$, captures the relative distance from node v to node u,

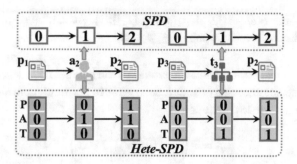

Fig. 4.4 An illustrative example of SPD and $Hete\text{-}SPD$ calculating $\rho_{p_1\rightsquigarrow p_2}$ and $\rho_{p_3\rightsquigarrow p_2}$. SPD does not consider the path type and assigns the same distance to them. While, $Hete\text{-}SPD$ calculates the relative distance with regard to node types, and is able to distinguish two different paths (i.e., $\mathbf{d}(p_2|p_1) = [1, 1, 0]$ and $\mathbf{d}(p_2|p_3) = [1, 0, 1]$)

$$spd(u|v) = \min \left\{ |\rho_{v\rightsquigarrow u}| \,\Big|\, \forall \rho_{v\rightsquigarrow u} \in \mathcal{P}_{v\rightsquigarrow u} \right\}, \tag{4.19}$$

where $|\rho_{v\rightsquigarrow u}|$ denotes the number of nodes in the path $\rho_{v\rightsquigarrow u}$ except the first node v. □

Example. As shown in Fig. 4.4, the shortest path $\rho_{p_1\rightsquigarrow p_2}$ from node p_1 to node p_2 is (p_1, a_2, p_2) and the corresponding $spd(p_2|p_1)$ is 2.

4.3.2.1 Heterogeneous Distance Encoding

To properly measure the relative distance between nodes on the HGs, heterogeneous shortest path distance ($Hete\text{-}SPD$) models both path length and path types, then encodes it as the node correlations via heterogeneous distance encoding.

Definition (*Heterogeneous Shortest Path Distance*) Given a node pair (u, v), heterogeneous shortest path distance describes the relative distance from node v to node u, denoted as $\mathbf{d}(u|v) \in \mathbb{R}^{|\mathcal{A}|}$, considering both the path length and the path type simultaneously. The jth dimension of $Hete\text{-}SPD$, which captures the relative distance with regard to node type $\mathcal{A}_j$, is shown as follows:

$$\mathbf{d}_j(u|v) = \min \left\{ |\rho_{v\rightsquigarrow u}|_{\phi(w)=j} \,\Big|\, \forall \rho_{v\rightsquigarrow u} \in \mathcal{P}_{v\rightsquigarrow u} \right\}, \tag{4.20}$$

where $|\rho_{v\rightsquigarrow u}|_{\phi(w)=j}$ denotes the number of nodes belonging to type-j in the path $\rho_{v\rightsquigarrow u}$ except the first node v. □

Example. As shown in Fig. 4.4a, the $Hete\text{-}SPD$ $\mathbf{d}(p_2|p_1)$ is $[1, 1, 0]$, indicating that the heterogeneous shortest path from node p_1 to node p_2 goes through one author (a_2) and one paper (p_2). Also, the *Hete-SPD* $\mathbf{d}(p_2|p_3)$ is $[1, 0, 1]$. As can be seen, *Hete-SPD* will assign different distances to node pairs (p_3, p_2) and (p_1, p_2).

Definition (*Heterogeneous Distance Encoding*) Given a target node set $S \subset \mathcal{V}$ in HG, the heterogeneous distance encoding of node i, denoted as $\mathbf{h}_i^S$, is the combination of *Hete-SPD* from all nodes in S to node i, as follows:

$$\mathbf{h}_i^S = F\Big(Enc\big(\mathbf{d}(i|v)\big) \Big| v \in S \Big), \tag{4.21}$$

where F is the fusion function, Enc is the encoding function.   □

Example. Given a node pair (u, v) remaining to predict (i.e., $S = \{u, v\}$), the HDE of node i with regard to node pair $\{u, v\}$ is shown as follows:

$$\mathbf{h}_i^{\{u,v\}} = F\Big(Enc\big(\mathbf{d}(i|u)\big), Enc\big(\mathbf{d}(i|v)\big) \Big). \tag{4.22}$$

Specifically, F is the concatenate operator and Enc is the concatenation of element-wise one-hot encoding, as follows:

$$Enc\big(\mathbf{d}(i|u)\big) = \Big\|_j onehot\Big(\min\big(\mathbf{d}_j(i|u), d_j^{\max}\big)\Big), \tag{4.23}$$

where $d_j^{\max}$ is the maximum distance for node type $\mathcal{A}_j$.

4.3.2.2 Heterogeneous Distance Encoding for Link Prediction

Link prediction is taken as an evaluation task to verify the effectiveness of HDE. The basic idea of DHN is to capture the correlations between nodes and integrate such correlations into the aggregating process of HGNN. Specifically, given a node pair (u, v), DHN calculates the HDE to capture the correlation between them and initialize neighbor embeddings. After that, DHN aggregates the heterogeneous neighbor's embeddings and injects the correlation into the final node embedding. Lastly, the final embedding of the node pair is used to predict the probability of link (u, v) existing.

Node Embedding Initialization. Significantly different from previous works [200, 233], DHN initializes node embeddings via HDE to capture the correlations between nodes. Specifically, given the link (u, v), DHN utilizes both (u, v)-specific HDE and the heterogeneous type encoding to initialize embedding.

    Heterogeneous distance encoding. Given a node pair (u, v), HDE is first calculated from node pair (u, v). Practically, k-hop heterogeneous enclosing subgraph $\mathcal{G}_{\{u,v\}}^k$ of node pair is extracted and only the HDE for node $i \in \mathcal{G}_{\{u,v\}}^k$ is calculated.

Heterogeneous type encoding. Heterogeneous type encoding is further utilized to initialize node embedding,

$$\mathbf{c}_i = onehot(\phi(i)), \tag{4.24}$$

where $\mathbf{c}_i \in \mathbb{R}^{|\mathcal{A}|}$ indicates the type of node i. As shown in Fig. 4.3, p_1's type index is $j = \phi(p_1) = 0$ and the corresponding heterogeneous type encoding is $[1, 0, 0]$. Lastly, heterogeneous type encoding and heterogeneous distance encoding of node i are concatenated and projected via MLP as the initial embedding to predict the existence of link (u, v), as follows:

$$\mathbf{e}_i^{\{u,v\}} = \sigma(\mathbf{W}_0 \cdot \mathbf{c}_i || \mathbf{h}_i^{\{u,v\}} + \mathbf{b}_0), \tag{4.25}$$

where $\mathbf{e}_i^{\{u,v\}}$ denotes the (u, v)-specific initial embedding of node i, $\mathbf{W}_0$ and $\mathbf{b}_0$ denote the weight matrix and the bias vector, respectively.

Heterogeneous Graph Convolution. After initializing the (u, v)-specific node embedding, heterogeneous graph convolution is further designed to aggregate neighbors in the heterogeneous enclosing subgraph. Taking node u as an example, a fixed number of neighbors $\mathcal{N}_u^{\{u,v\}}$ are first sampled, and then an extremely simple aggregating function is used as follows:

$$\mathbf{x}_{u,l}^{\{u,v\}} = \sigma(\mathbf{W}^l \cdot (\mathbf{x}_{u,l-1}^{\{u,v\}} || Avg\{\mathbf{x}_{i,l-1}^{\{u,v\}}\} + \mathbf{b}^l), \forall i \in \mathcal{N}_u^{\{u,v\}}, \tag{4.26}$$

where $\mathbf{W}^l$ and $\mathbf{b}^l$ denote the weight matrix and the bias vector, respectively. After L-layer aggregation, the final node embedding $\mathbf{z}_u^{\{u,v\}} = \mathbf{x}_{u,L}^{\{u,v\}}$ is obtained.

Loss Function and Optimization. Given a (u, v)-specific embedding of node pair (i.e., $\mathbf{z}_u^{\{u,v\}}$ and $\mathbf{z}_v^{\{u,v\}}$), the concatenation of them is fed into a MLP to get the predicting score $\hat{y}_{u,v}$ as follows:

$$\hat{y}_{u,v} = \sigma\left(\mathbf{W}_1 \cdot (\mathbf{z}_u^{\{u,v\}} || \mathbf{z}_v^{\{u,v\}}) + b_1\right), \tag{4.27}$$

where $\mathbf{W}_1$ and b_1 denote the weight vector and the bias scalar, respectively. The loss function is shown as follows:

$$\mathcal{L} = \sum_{(u,v) \in \mathcal{E}^+ \cup \mathcal{E}^-} \left(y_{u,v} \log \hat{y}_{u,v} + \left(1 - y_{u,v}\right) \log\left(1 - \hat{y}_{u,v}\right)\right), \tag{4.28}$$

where $\mathcal{E}$ and $\mathcal{E}^-$ denote positive and negative node pairs, respectively.

4.3.3   Experiments

4.3.3.1 Datasets and Baselines

As shown in Table 4.3, experiments are conducted on three real-world heterogeneous graphs and compare DHN with state-of-the-art baselines including homogeneous GNNs (i.e., GCN [102], GAT [189], SAGE [74], GIN [215], and DEGNN [115]) and HGNNs (i.e., HAN [200], MEIRec [45], and HGT [84]).

Table 4.3 Statistics of the datasets

Datasets	A-B	#A	#B	#A-B	#$\mathcal{V}$	#$\mathcal{E}$
LastFM	Artist-Tag	1181	539	1500	3790	4500
	User-Artist	1496	1755	3000		
ACM	Term-Paper	381	769	1500	3908	4500
	Paper-Author	1000	2527	3000		
IMDB	Movie-Actor	3061	1374	7071	5296	10132
	Movie-Director	3061	861	3061		
FreeBase	Movie-Writer	3492	4459	6414	43854	75517
	Movie-Director	3492	2502	3762		
	Movie-Actor	3492	33401	65341		

Table 4.4 Quantitative results (%) on the transductive (T) and inductive (I) link prediction

Setting	Models	LastFM				ACM				IMDB			
		A-T		U-A		P-A		T-P		M-A		M-D	
		Acc	AUC	Acc	AUC	Acc	AUC	Acc	AUC	Acc	AUC	Acc	AUC
T	GCN	50.12	50.31	50.21	50.41	50.12	50.43	50.31	50.87	50.51	50.98	50.89	51.02
	SAGE	73.43	83.24	67.66	73.61	77.41	77.81	65.12	78.12	54.23	63.23	84.82	89.01
	GAT	50.54	50.61	51.02	51.21	50.12	50.67	51.21	51.98	51.01	51.44	51.23	51.92
	GIN	84.71	93.06	72.81	79.19	77.28	83.72	78.33	82.32	73.62	80.45	84.83	97.42
	DEGNN	85.32	93.72	73.33	80.49	85.12	91.79	79.78	85.51	78.57	88.28	94.18	96.26
	HAN	81.54	87.32	71.12	78.12	84.23	88.77	81.23	89.31	79.49	87.99	94.77	95.84
	MEIRec	84.27	93.24	72.21	86.32	84.08	90.18	81.51	89.86	79.06	85.23	95.75	96.64
	HGT	86.12	94.12	74.12	81.95	86.12	92.32	82.68	90.99	81.21	91.21	96.77	97.23
	DHN	**95.21**	**97.43**	**96.23**	**97.81**	**95.23**	**98.32**	**85.31**	**92.81**	**83.33**	**94.22**	**99.12**	**99.31**
	Imp.	10.6	3.5	29.8	13.3	10.6	6.5	3.2	2.1	2.6	3.3	2.4	2.4
I	GCN	50.21	50.91	50.31	50.89	51.21	51.52	50.12	53.2	50.89	51.02	51.31	52.34
	SAGE	63.11	59.45	66.8	70.29	57.22	52.12	52.12	54.44	61.32	62.31	63.12	64.21
	GAT	51.21	52.12	50.91	51.02	52.13	52.98	51.32	52.73	51.02	53.01	52.02	53.52
	GIN	80.33	82.53	66.83	71.16	60.23	62.31	61.71	61.57	55.12	64.12	70.41	75.01
	DEGNN	81.97	87.49	66.12	72.21	72.21	80.13	63.28	70.72	67.23	71.22	79.21	81.21
	HAN	79.51	85.12	78.85	83.99	81.91	85.23	78.91	86.89	80.61	86.15	90.22	93.12
	MEIRec	80.33	87.62	80.28	83.87	83.33	90.19	77.32	85.33	80.22	86.14	93.73	95.31
	HGT	83.44	89.22	83.01	86.21	85.12	92.13	79.22	87.53	81.12	89.21	95.21	96.52
	DHN	**88.67**	**94.88**	**92.68**	**98.15**	**91.22**	**95.07**	**82.05**	**94.85**	**82.95**	**94.46**	**98.19**	**99.51**
	Imp.	6.2	6.3	11.6	13.8	7.1	3.2	3.5	8.3	2.2	5.8	3.1	3

4.3.3.2 Link Prediction

Link prediction has been widely used to test the generalization ability of graph neural networks. As shown in Table 4.4, both transductive and inductive link prediction are tested. The major findings are shown as follows:

- DHN consistently performs better than all baselines with significant improvements on all datasets in both transductive and inductive settings. It demonstrates the effectiveness of heterogeneous distance encoding in modeling node pair correlation.
- Compared to the traditional GNNs which individually learn node embedding (e.g., GCN, GraphSAGE, GIN, HAN, MEIRec, and HGT), although both DEGNN and DHN capture the correlations between nodes via distance modeling, $Hete\text{-}SPD$ is more powerful than SPD because it fully considers both path length and path type.
- With the full consideration of graph heterogeneity, HGNNs including HAN, MEIRec, HGT, and DHN usually show superiorities over homogeneous GNNs. When predicting heterogeneous links, node type encoding may provide potentially valuable information and improves the performance.

4.3.3.3 Heterogeneous Distance Encoding Analysis

The correlation modeling via HDE plays a key role in improving the representational power of HGNNs. As shown in Fig. 4.5, the characteristic of HDE (w.r.t. maximum distance and dimension) is observed as follows: From Fig. 4.5a, it is found that the relative distance in a small range (typically 2-3) provides valuable correlation information. Comparing Fig. 4.5a and Table 4.4, even with $d^{max} = 1$, DHN performs far better than all the baselines. It makes sense because the local structure around nodes is beneficial for link prediction [234] while long-range connections may introduce noise and lead to over-fitting. As shown in Fig. 4.5b, with the growth of the dimension, the effectiveness of HDE rises first and then starts to keep stable, which means it needs enough dimension to encode the correlation between nodes, and larger dimension may introduce additional redundancies.

The more detailed description and experiment validation can be seen in [89].

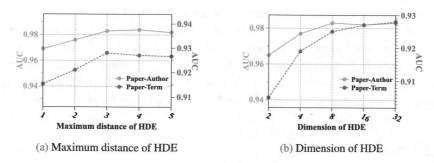

(a) Maximum distance of HDE   (b) Dimension of HDE

Fig. 4.5 Effectiveness of HDE w.r.t. maximum distance and dimension

4.4   Self-supervised HGNN with Co-contrastive Learning

4.4.1   Overview

Contrastive learning, as one typical technique of self-supervised learning, has attracted considerable attention [24, 77, 78, 188]. By maximizing the similarity between positive samples while minimizing the similarity between negative samples, contrastive learning is able to learn the discriminative embeddings even without labels. Despite the wide use of contrastive learning in computer vision [24, 78] and natural language processing [40, 108], little effort has been made toward investigating the great potential on HGs.

In practice, designing HGNNs with contrastive learning is non-trivial, which requires addressing the following three fundamental problems. (i) *How to design a heterogeneous contrastive mechanism.* A HG consists of multiple types of nodes and relations, which naturally implies it possesses very complex structures. To learn an effective node embedding that can fully encode these semantics, performing contrastive learning only on a single meta-path view [149] is actually distant from sufficient. (ii) *How to select proper views in a HG.* Despite that one can extract many different views from a HG because of the heterogeneity, one fundamental requirement is that the selected views should cover both of the local and high-order structures. (iii) *How to set a difficult contrastive task.* It is well known that a proper contrastive task will further promote learning a more discriminative embedding [24, 78, 184]. If two views are too similar, the supervised signal will be too weak to learn informative embedding. So the contrastive learning on these two views need to be more complicated.

Therefore, self-supervised Heterogeneous graph neural network with Co-contrastive learning (HeCo) studies the problem of self-supervised learning on HGs. Specifically, different from previous contrastive learning which contrasts the original network and the corrupted network, HeCo chooses network schema and meta-path structure as two views to collaboratively supervise each other. The network schema view is able to capture the local structure, while the meta-path view aims at capturing high-order structure. To make contrast harder, a view mask mechanism hides different parts of network schema and meta-path, respectively, which will further enhance the diversity of views and help extract higher level factors. Finally, HeCo modestly adapts traditional contrastive loss to the graph data, where a node has many positive samples rather than only one [24, 78]. With the training going on, these two views are guided by each other and collaboratively optimize.

4.4.2   The HeCo Method

The overall architecture of HeCo is shown in Fig. 4.6. It encodes nodes from network schema view and meta-path view. During the encoding, HeCo creatively involves a view mask mechanism, which makes these two views complement and supervise mutually. With the two view-specific embeddings, a contrastive learning is employed across these two views.

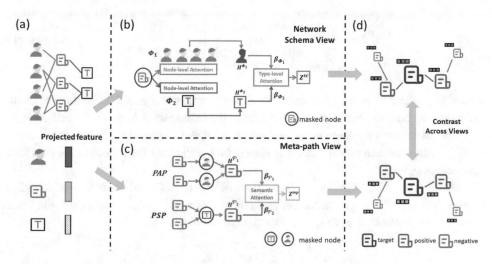

Fig. 4.6 The overall architecture of HeCo

4.4.2.1 Node Feature Transformation

Because there are different types of nodes in a HG, features of all types of nodes need to be projected into a common latent vector space, as shown in Fig. 6.1a. Specifically, for a node i with type ϕ_i, a type-specific mapping matrix W_{ϕ_i} is designed to transform its feature x_i into common space as follows:

$$h_i = \sigma \left(W_{\phi_i} \cdot x_i + b_{\phi_i} \right), \tag{4.29}$$

where $h_i \in \mathbb{R}^{d \times 1}$ is the projected feature of node i, $\sigma(\cdot)$ is an activation function, and b_{ϕ_i} denotes as vector bias, respectively.

4.4.2.2 Network Schema View Guided Encoder

According to network schema, assuming that the target node i connects with S other types of nodes$\{\Phi_1, \Phi_2, \ldots, \Phi_S\}$, the neighbors with type Φ_m of node i can be defined as $N_i^{\Phi_m}$. For node i, different types of neighbors contribute differently to its embedding, so attention mechanism is employed here at node-level and type-level to hierarchically aggregate messages. Specifically, node-level attention is first applied to fuse neighbors with type Φ_m:

$$h_i^{\Phi_m} = \sigma \left(\sum_{j \in N_i^{\Phi_m}} \alpha_{i,j}^{\Phi_m} \cdot h_j \right), \tag{4.30}$$

where σ is a non-linear activation, h_j is the projected feature of node j. And $\alpha_{i,j}^{\Phi_m}$ denotes the attention value of node j with type Φ_m to node i, calculated as follows:

$$\alpha_{i,j}^{\Phi_m} = \frac{\exp\left(Leaky ReLU\left(\mathbf{a}_{\Phi_m}^{\top} \cdot [h_i \| h_j]\right)\right)}{\sum\limits_{l \in N_i^{\Phi_m}} \exp\left(Leaky ReLU\left(\mathbf{a}_{\Phi_m}^{\top} \cdot [h_i \| h_l]\right)\right)}, \tag{4.31}$$

where $\mathbf{a}_{\Phi_m} \in \mathbb{R}^{2d \times 1}$ is the node-level attention vector for Φ_m and $\|$ denotes concatenate operation. In practice, HeCo does not aggregate the information from all the neighbors in $N_i^{\Phi_m}$, but randomly samples a part of neighbors every epoch. In this way, it promotes diversity of embeddings in each epoch under this view, which will make the following contrast task more challenging.

Once all type embeddings $\{h_i^{\Phi_1}, \ldots, h_i^{\Phi_S}\}$ are obtained, HeCo utilizes type-level attention to fuse them together to get the final embedding z_i^{sc} for node i under network schema view. First, the weight of each node type is measured as follows:

$$w_{\Phi_m} = \frac{1}{|V|} \sum_{i \in V} \mathbf{a}_{sc}^{\top} \cdot \tanh\left(\mathbf{W}_{sc} h_i^{\Phi_m} + \mathbf{b}_{sc}\right),$$

$$\beta_{\Phi_m} = \frac{\exp\left(w_{\Phi_m}\right)}{\sum_{i=1}^{S} \exp\left(w_{\Phi_i}\right)}, \tag{4.32}$$

where V is the set of target nodes, $\mathbf{W}_{sc} \in \mathbb{R}^{d \times d}$ and $\mathbf{b}_{sc} \in \mathbb{R}^{d \times 1}$ are learnable parameters, and $\mathbf{a}_{sc}$ denotes type-level attention vector. β_{Φ_m} is interpreted as the importance of type Φ_m to target node i. So, the type embeddings are weighted sum to get z_i^{sc}:

$$z_i^{sc} = \sum_{m=1}^{S} \beta_{\Phi_m} \cdot h_i^{\Phi_m}. \tag{4.33}$$

4.4.2.3 Meta-Path View Guided Encoder

Given a meta-path $\mathcal{P}_n$ from M meta-paths $\{\mathcal{P}_1, \mathcal{P}_2, \ldots, \mathcal{P}_M\}$ that start from node i, the meta-path-based neighbors are denoted as $N_i^{\mathcal{P}_n}$. For example, as shown in Fig. 4.7a, P_2 is a neighbor of P_3 based on meta-path PAP. Each meta-path represents one semantic similarity, and meta-path specific GCN [102] is applied to encode this characteristic:

$$h_i^{\mathcal{P}_n} = \frac{1}{d_i + 1} h_i + \sum_{j \in N_i^{\mathcal{P}_n}} \frac{1}{\sqrt{(d_i + 1)(d_j + 1)}} h_j, \tag{4.34}$$

where d_i and d_j are degrees of node i and j, and h_i and h_j are their projected features, respectively. With M meta-paths, M embeddings $\{h_i^{\mathcal{P}_1}, \ldots, h_i^{\mathcal{P}_M}\}$ are obtained for node i. Then semantic-level attention is utilized to fuse them into the final embedding z_i^{mp} under the meta-path view:

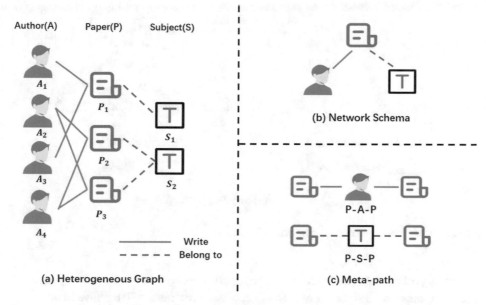

Fig. 4.7 A toy example of HG (ACM) and relative illustrations of meta-path and network schema

$$z_i^{mp} = \sum_{n=1}^{M} \beta_{\mathcal{P}_n} \cdot h_i^{\mathcal{P}_n}, \qquad (4.35)$$

where $\beta_{\mathcal{P}_n}$ weighs the importance of meta-path $\mathcal{P}_n$, which is calculated as follows:

$$w_{\mathcal{P}_n} = \frac{1}{|V|} \sum_{i \in V} \mathbf{a}_{mp}^\top \cdot \tanh\left(\mathbf{W}_{mp} h_i^{\mathcal{P}_n} + \mathbf{b}_{mp}\right),$$

$$\beta_{\mathcal{P}_n} = \frac{\exp\left(w_{\mathcal{P}_n}\right)}{\sum_{i=1}^{M} \exp\left(w_{\mathcal{P}_i}\right)}, \qquad (4.36)$$

where $\mathbf{W}_{mp} \in \mathbb{R}^{d \times d}$ and $\mathbf{b}_{mp} \in \mathbb{R}^{d \times 1}$ are the learnable parameters, and $\mathbf{a}_{mp}$ denotes the semantic-level attention vector.

4.4.2.4 View Mask Mechanism

During the generation of z_i^{sc} and z_i^{mp}, a view mask mechanism is designed to hide different parts of network schema and meta-path views, respectively. In particular, a schematic diagram on ACM is given in Fig. 4.8, where the target node is P_1. In the process of network schema encoding, P_1 only aggregates its neighbors including authors A_1, A_2 and subject S_1 into z_1^{sc}, but the message from itself is masked. While in the process of meta-path encoding, message only passes along meta-paths (e.g., PAP, PSP) from P_2 and P_3 to target P_1 to generate z_1^{mp}, while the information of intermediate nodes A_1 and S_1 are discarded. Therefore, the

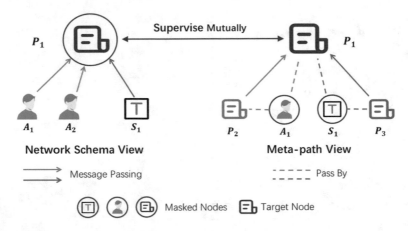

Fig. 4.8 A schematic diagram of view mask mechanism

embeddings of node P_1 learned from these two parts are correlated but also complementary. They can supervise the training of each other, which presents a collaborative trend.

4.4.2.5 Collaboratively Contrastive Optimization

After getting the z_i^{sc} and z_i^{mp} for node i from the above two views, they are fed into a MLP to map them into the space where contrastive loss is calculated:

$$
\begin{aligned}
z_i^{sc}\_proj &= W^{(2)}\sigma\left(W^{(1)}z_i^{sc} + b^{(1)}\right) + b^{(2)}, \\
z_i^{mp}\_proj &= W^{(2)}\sigma\left(W^{(1)}z_i^{mp} + b^{(1)}\right) + b^{(2)},
\end{aligned}
\tag{4.37}
$$

where σ is ELU non-linear function. Next, when calculating contrastive loss, positive and negative samples need to be defined in HG. Given a node under network schema view, a new positive selection strategy is proposed, i.e., if two nodes are connected by many meta-paths, they are positive samples, as shown in Fig. 6.1d where links between papers represent they are positive samples of each other. One advantage of such strategy is that the selected positive samples can well reflect local structure of the target node.

For node i and j, a function $\mathbb{C}_i(\cdot)$ is defined to count the number of meta-paths connecting these two nodes:

$$
\mathbb{C}_i(j) = \sum_{n=1}^{M} \mathbb{K}\left(j \in N_i^{\mathcal{P}_n}\right),
\tag{4.38}
$$

where $\mathbb{K}(\cdot)$ is the indicator function. Then HeCo constructs a set $S_i = \{j \mid j \in V \text{ and } \mathbb{C}_i(j) \neq 0\}$ and sorts it in the descending order based on the value of $\mathbb{C}_i(\cdot)$. Next if $|S_i| > T_{pos}$, where T_{pos} is a threshold, first T_{pos} nodes from S_i are selected as positive samples of i, denotes as

$\mathbb{P}_i$, otherwise all nodes in S_i are retained. And all left nodes are treated as negative samples of i, denotes as $\mathbb{N}_i$.

With the positive sample set $\mathbb{P}_i$ and negative sample set $\mathbb{N}_i$, the contrastive loss under network schema view is calculated:

$$\mathcal{L}_i^{sc} = -\log \frac{\sum_{j \in \mathbb{P}_i} exp\left(sim\left(z_i^{sc}\_proj, z_j^{mp}\_proj\right)/\tau\right)}{\sum_{k \in \{\mathbb{P}_i \bigcup \mathbb{N}_i\}} exp\left(sim\left(z_i^{sc}\_proj, z_k^{mp}\_proj\right)/\tau\right)}, \tag{4.39}$$

where $sim(u, v)$ denotes the cosine similarity between two vectors u and v, and τ denotes a temperature parameter. Please note that for two nodes in a pair, the target embedding is from the network schema view ($z_i^{sc}\_proj$) and the embeddings of positive and negative samples are from the meta-path view ($z_k^{mp}\_proj$). In this way, HeCo realizes the cross-view self-supervision.

The contrastive loss $\mathcal{L}_i^{mp}$ is similar as $\mathcal{L}_i^{sc}$, but differently, the target embedding is from the meta-path view while the embeddings of positive and negative samples are from the network schema view. The overall objective is given as follows:

$$\mathcal{J} = \frac{1}{|V|} \sum_{i \in V} \left[\lambda \cdot \mathcal{L}_i^{sc} + (1 - \lambda) \cdot \mathcal{L}_i^{mp}\right], \tag{4.40}$$

where λ is a coefficient to balance the effect of two views. In the end, z^{mp} is used to perform downstream tasks because nodes of target type explicitly participate into the generation of z^{mp}.

4.4.3   Experiments

4.4.3.1 Experimental Setup
Datasets. As shown in Table 4.5, experiments are conducted on four real-world heterogeneous graphs, i.e., ACM [248], DBLP [56], Freebase [118] and AMiner [82].
Baselines. HeCo is compared with three categories of baselines: unsupervised homogeneous methods (i.e., GraphSAGE [74] and DGI [188]), unsupervised heterogeneous methods (i.e., HERec [166], HetGNN [233] and DMGI [149]), and a semi-supervised heterogeneous method HAN [200].

4.4.3.2 Node Classification
The learned embeddings of nodes are used to train a linear classifier. To more comprehensively evaluate HeCo, 20, 40, 60 labeled nodes per class are chosen as training set, and 1000 nodes are selected as validation and 1000 as test set respectively, for each dataset. Common evaluation metrics are used, including Macro-F1, Micro-F1, and AUC. The results are reported in Table 4.6.

Table 4.5 The statistics of the datasets

Dataset	Node	Relation	Meta-path
ACM	paper (P):4019 author (A):7167 subject (S):60	P-A:13407 P-S:4019	PAP PSP
DBLP	author (A):4057 paper (P):14328 conference (C):20 term (T):7723	P-A:19645 P-C:14328 P-T:85810	APA APCPA APTPA
Freebase	movie (M):3492 actor (A):33401 direct (D):2502 writer (W):4459	M-A:65341 M-D:3762 M-W:6414	MAM MDM MWM
AMiner	paper (P):6564 author (A):13329 reference (R):35890	P-A:18007 P-R:58831	PAP PRP

As can be seen, HeCo generally outperforms all the other methods on all datasets and all splits, even compared with HAN, a semi-supervised method. It can also be observed that HeCo outperforms DMGI in most cases, while DMGI is even worse than other baselines with some settings, indicating that single-view is noisy and incomplete. So, performing contrastive learning across views is effective. Moreover, even HAN utilizes the label information, HeCo performs better than it in all cases. This well indicates the great potential of cross-view contrastive learning.

4.4.3.3 Collaborative Trend Analysis

One salient property of HeCo is the cross-view collaborative mechanism, i.e., HeCo employs the network schema and meta-path views to collaboratively learn the embeddings. This section examines the changing trends of type-level attention β_Φ in network schema view and semantic-level attention $\beta_\mathcal{P}$ in meta-path view w.r.t. epochs, and the results are plotted in Fig. 4.9. For both ACM and AMiner, the changing trends of two views are collaborative and consistent. Specifically, for ACM, β_Φ of type A is higher than type S, and $\beta_\mathcal{P}$ of meta-path PAP also exceeds that of PSP. For AMiner, type R and meta-path PRP are more important in two views, respectively. This indicates that network schema view and meta-path view adapt for each other during training and collaboratively optimize by contrasting each other.

The more detailed description and experiment validation can be seen in [201].

Table 4.6 Quantitative results ($\%\pm\sigma$) on node classification

Datasets	Metric	Split	GraphSAGE	HERec	HetGNN	HAN	DGI	DMGI	HeCo
ACM	Ma-F1	20	47.13±4.7	55.13±1.5	72.11±0.9	85.66±2.1	79.27±3.8	87.86±0.2	**88.56±0.8**
		40	55.96±6.8	61.61±3.2	72.02±0.4	87.47±1.1	80.23±3.3	86.23±0.8	**87.61±0.5**
		60	56.59±5.7	64.35±0.8	74.33±0.6	88.41±1.1	80.03±3.3	87.97±0.4	**89.04±0.5**
	Mi-F1	20	49.72±5.5	57.47±1.5	71.89±1.1	85.11±2.2	79.63±3.5	87.60±0.8	**88.13±0.8**
		40	60.98±3.5	62.62±0.9	74.46±0.8	87.21±1.2	80.41±3.0	86.02±0.9	**87.45±0.5**
		60	60.72±4.3	65.15±0.9	76.08±0.7	88.10±1.2	80.15±3.2	87.82±0.5	**88.71±0.5**
	AUC	20	65.88±3.7	75.44±1.3	84.36±1.0	93.47±1.5	91.47±2.3	**96.72±0.3**	96.49±0.3
		40	71.06±5.2	79.84±0.5	85.01±0.6	94.84±0.9	91.52±2.3	96.35±0.3	**96.40±0.4**
		60	70.45±6.2	81.64±0.7	87.64±0.7	94.68±1.4	91.41±1.9	**96.79±0.2**	96.55±0.3
DBLP	Ma-F1	20	71.97±8.4	89.57±0.4	89.51±1.1	89.31±0.9	87.93±2.4	89.94±0.4	**91.28±0.2**
		40	73.69±8.4	89.73±0.4	88.61±0.8	88.87±1.0	88.62±0.6	89.25±0.4	**90.34±0.3**
		60	73.86±8.1	90.18±0.3	89.56±0.5	89.20±0.8	89.19±0.9	89.46±0.6	**90.64±0.3**
	Mi-F1	20	71.44±8.7	90.24±0.4	90.11±1.0	90.16±0.9	88.72±2.6	90.78±0.3	**91.97±0.2**
		40	73.61±8.6	90.15±0.4	89.03±0.7	89.47±0.9	89.22±0.5	89.92±0.4	**90.76±0.3**
		60	74.05±8.3	91.01±0.3	90.43±0.6	90.34±0.8	90.35±0.8	90.66±0.5	**91.59±0.2**
	AUC	20	90.59±4.3	98.21±0.2	97.96±0.4	98.07±0.6	96.99±1.4	97.75±0.3	**98.32±0.1**
		40	91.42±4.0	97.93±0.1	97.70±0.3	97.48±0.6	97.12±0.4	97.23±0.2	**98.06±0.1**
		60	91.73±3.8	98.49±0.1	97.97±0.2	97.96±0.5	97.76±0.5	97.72±0.4	**98.59±0.1**
Freebase	Ma-F1	20	45.14±4.5	55.78±0.5	52.72±1.0	53.16±2.8	54.90±0.7	55.79±0.9	**59.23±0.7**
		40	44.88±4.1	59.28±0.6	48.57±0.5	59.63±2.3	53.40±1.4	49.88±1.9	**61.19±0.6**
		60	45.16±3.1	56.50±0.4	52.37±0.8	56.77±1.7	53.81±1.1	52.10±0.7	**60.13±1.3**
	Mi-F1	20	54.83±3.0	57.92±0.5	56.85±0.9	57.24±3.2	58.16±0.9	58.26±0.9	**61.72±0.6**
		40	57.08±3.2	62.71±0.7	53.96±1.1	63.74±2.7	57.82±0.8	54.28±1.6	**64.03±0.7**
		60	55.92±3.2	58.57±0.5	56.84±0.7	61.06±2.0	57.96±0.7	56.69±1.2	**63.61±1.6**
	AUC	20	67.63±5.0	73.89±0.4	70.84±0.7	73.26±2.1	72.80±0.6	73.19±1.2	**76.22±0.8**
		40	66.42±4.7	76.08±0.4	69.48±0.2	77.74±1.2	72.97±1.1	70.77±1.6	**78.44±0.5**
		60	66.78±3.5	74.89±0.4	71.01±0.5	75.69±1.5	73.32±0.9	73.17±1.4	**78.04±0.4**
AMiner	Ma-F1	20	42.46±2.5	58.32±1.1	50.06±0.9	56.07±3.2	51.61±3.2	59.50±2.1	**71.38±1.1**
		40	45.77±1.5	64.50±0.7	58.97±0.9	63.85±1.5	54.72±2.6	61.92±2.1	**73.75±0.5**
		60	44.91±2.0	65.53±0.7	57.34±1.4	62.02±1.2	55.45±2.4	61.15±2.5	**75.80±1.8**
	Mi-F1	20	49.68±3.1	63.64±1.1	61.49±2.5	68.86±4.6	62.39±3.9	63.93±3.3	**78.81±1.3**
		40	52.10±2.2	71.57±0.7	68.47±2.2	76.89±1.6	63.87±2.9	63.60±2.5	**80.53±0.7**
		60	51.36±2.2	69.76±0.8	65.61±2.2	74.73±1.4	63.10±3.0	62.51±2.6	**82.46±1.4**
	AUC	20	70.86±2.5	83.35±0.5	77.96±1.4	78.92±2.3	75.89±2.2	85.34±0.9	**90.82±0.6**
		40	74.44±1.3	88.70±0.4	83.14±1.6	80.72±2.1	77.86±2.1	88.02±1.3	**92.11±0.6**
		60	74.16±1.3	87.74±0.5	84.77±0.9	80.39±1.5	77.21±1.4	86.20±1.7	**92.40±0.7**

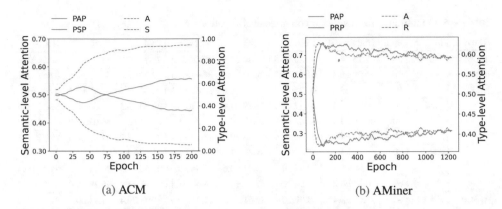

Fig. 4.9 The collaborative changing trends of attentions in two views

4.5   Conclusion

Due to the capacity of modeling various types of nodes and diverse interactions between them, heterogeneous graphs (HGs) are ubiquitous. Recently, heterogeneous graph neural networks (HGNNs) are proposed to combine the mechanism of message passing with heterogeneity, so that the complex structures and rich semantics in HGs can be well preserved.

However, HGNNs are not perfect but have many inherent limitations. In this chapter, we introduce three representative HGNNs that focus on the deep degradation phenomenon or limited discriminative power. Specifically, we firstly introduce a heterogeneous graph propagation network (HPN) to alleviate the semantic confusion faced by the growth of model depth. Next, we introduce a distance encoding-based heterogeneous graph neural network (DHN), which injects the heterogeneous shortest path distance into the aggregating process to learn more representative node embeddings. Finally, we introduce a self-supervised Heterogeneous graph neural network with Co-contrastive learning (HeCo) to arm HGNNs with cross-view contrastive learning. Experiments verify the effectiveness of these methods on various tasks and datasets.

4.6   Further Reading

Heterogeneous graphs (HGs) have become ubiquitous in real-world scenarios, ranging from bibliographic networks, social networks to recommendation systems. It has been well recognized that HG is a powerful model to embrace rich semantics and structural information. Therefore, researches on HG data have been experiencing tremendous growth in data mining and machine learning.

Different from graph neural networks (GNNs) that can directly fuse the attributes of neighbors to update node embeddings, HGNNs need to overcome the heterogeneity of attributes and design effective fusion methods to utilize the neighborhood information. Fortunately, HGNNs have made great progress in recent years, which clearly shows that it is a powerful and promising graph analysis paradigm. In addition to the three methods described in this chapter, some typical methods are briefly summarized below. Wang et al. [200] and Zhang et al. [233] both leverage hierarchical aggregation to capture rich semantics. MEIRec [45] and IntentGC [249] apply HGNNs to solve the intent recommendation. Hu et al. [83] propose a heterogeneous graph attention network for short text classification. GTN [230] learns a soft selection of edge types and generates meta-paths automatically, solving the problem of meta-path selection. HGT [84] adopts heterogeneous mutual attention to aggregate meta-relation triplet, and MAGNN [56] leverages relational rotation encoder to aggregate meta-path instances.

We please refer the readers to the recent survey [197] for a more comprehensive review.

Dynamic Graph Neural Networks

5

Yugang Ji

5.1 Introduction

Graph neural networks have shed a light on network analysis due to their capability of encoding the structures and properties of networks with latent representations [15, 32]. The state-of-the-art [70, 154, 155, 181, 191] have achieved promising performance in many data mining tasks. Most of them focus on static networks with fixed structures. However, graphs in real-world scenarios usually exhibit complex temporal properties and heterogeneous types of edges. How to improve the representation learning of temporal heterogeneous graphs is still an open problem.

To capture the temporal evolution of dynamic graphs, it is general to split the whole graph into several snapshots and generate representations by inputting all snapshot-based embeddings into sequential models like long-short term memory (LSTM) and gated recurrent units (GRU) [67, 125, 147, 209, 260]. However, the temporal network evolution usually follows two dynamics processes, i.e., the microscopic dynamics of edge establishments and macroscopic dynamics of network scales. Besides, such design of dynamic GNNs is not suitable for modeling heterogeneous graphs because of overlooking the rich semantics within different-typed interactions.

In this chapter, we will introduce three newest works to model the complex dynamics of homogeneous and heterogeneous dynamic graphs. In Sect. 5.2, we will introduce a temporal graph embedding method (named M^2DNE) which models both micro- and macro-dynamics of edges based on temporal point process. In Sect. 5.3, we will introduce the dynamic heterogeneous graph embedding model (named HPGE) which preserves both semantics

Y. Ji (✉)
Beijing University of Posts and Telecommunications, Beijing, China

© The Author(s), under exclusive license to Springer Nature Switzerland AG 2023   87
C. Shi et al., *Advances in Graph Neural Networks*, Synthesis Lectures
on Data Mining and Knowledge Discovery,
https://doi.org/10.1007/978-3-031-16174-2_5

and dynamics by learning the formation process of all heterogeneous temporal events. In Sect. 5.4, we will introduce the dynamic meta-path guided temporal heterogeneous graph modeling approach (named DyMGNN) which designs dynamic meta-path and heterogeneous mutual evolution attention mechanisms to effectively captures the dynamic semantics and model the mutual evolution of different semantics.

5.2   Micro- and Macro-dynamics

5.2.1   Overview

A temporal network (or called temporal graph) naturally represents the evolution of a network, including not only the fine-grained network structure but also the macroscopic network scale. At the microscopic level, the temporal network structure is driven by the establishments of edges, which is actually a sequence of chronological events involving two nodes. As shown in Fig. 5.1, an edge generated at time t is inevitably related to the historical neighbors before t, and different neighborhoods may have distinct influences. For instance, the influences of (v_3, v_2, t_i) and (v_4, v_2, t_i) on the event (v_3, v_4, t_j) should be larger than that of (v_3, v_1, t_i), since nodes v_2, v_3 and v_4 form a closed triad [86, 254]. More importantly, at the macroscopic level, another salient property of the temporal network is that the network scale evolves with obvious distributions over time, e.g., an S-shaped sigmoid curve [112] or a power-law-like pattern [231]. As shown in Fig. 5.1, when the network evolves over time, the edges are continuously being built and form the network structures at each timestamp.

In this section, we propose a novel temporal network embedding method with micro- and macro-dynamics, named M^2DNE. In particular, to model the chronological events of

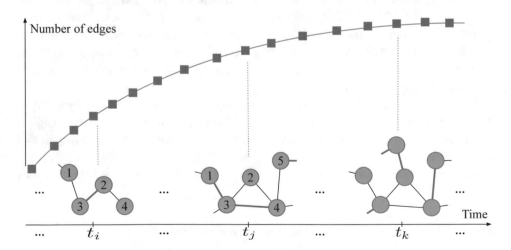

Fig. 5.1 A toy example of temporal network with micro- and macro-dynamics

edge establishments in a temporal network (i.e., micro-dynamics), we elaborately design a temporal attention point process by parameterizing the conditional intensity function with node embeddings, which captures the fine-grained structural and temporal properties with hierarchical temporal attention. To model the evolution pattern of the temporal network scale (i.e., macro-dynamics), we define a general dynamics equation as a non-linear function of the network embedding, which imposes constraints on the network embedding at a high structural level and well couples the analysis of the dynamics with representation learning on temporal networks. At last, we combine micro- and macro-dynamics preserved embedding and optimize them jointly. The proposed M^{2}DNE has the capability to capture the formation process of topological structures and the evolutionary pattern of network scale in a unified manner.

5.2.2 The M^{2}DNE Method

As illustrated in Fig. 5.2, from a microscopic perspective (i.e., Fig. 5.2a), we consider the establishment of edges as the chronological event and propose a temporal attention point process to capture the fine-grained structural and temporal properties for network embedding. The establishment of an edge (e.g., (v_3, v_4, t_j)) is determined by the nodes themselves and their historical neighbors (e.g., $\{v_1, v_2, \ldots\}$ and $\{v_5, v_2, \ldots\}$), where the distinct influences are captured with a hierarchical temporal attention. From a macroscopic perspective (i.e., Fig. 5.2b), the inherent evolution pattern of network scale constrains the network structures at a higher level, which is defined as a dynamics equation parameterized with network embedding $\mathbf{U}$ and timestamp t. Micro- and macro-dynamics evolve and derive node embeddings in a mutual manner (i.e., Fig. 5.2c).

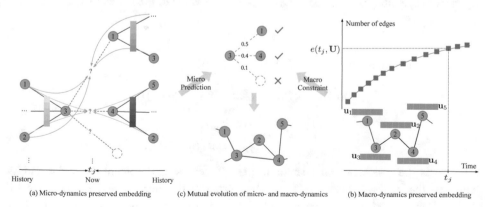

(a) Micro-dynamics preserved embedding   (c) Mutual evolution of micro- and macro-dynamics   (b) Macro-dynamics preserved embedding

Fig. 5.2 The overall architecture of M^{2}DNE

5.2.2.1 Micro-dynamics Preserved Embedding

Given a temporal edge $o = (i, j, t)$ (i.e., an observed event), we parameterize the intensity of the event $\tilde{\lambda}_{i,j}(t)$ with network embedding $\mathbf{U} = [\mathbf{u}_i]^\top$. Since similar nodes i and j are more likely to establish the edge (i, j, t), the similarity between nodes i and j should be proportional to the intensity of the event that i and j build a link at time t. Besides, the similarity between the historical neighbors and the current node indicates the degree of past impact on the event (i, j, t), which should decrease with time and be different from distinct neighbors. The occurrence intensity of the event $o = (i, j, t)$ consists of the base intensity from nodes themselves and the historical influences from two-way neighbors, as follows:

$$\tilde{\lambda}_{i,j}(t) = \underbrace{g(\mathbf{u}_i, \mathbf{u}_j)}_{Base\ Intensity} \tag{5.1}$$

$$+ \beta_{ij} \sum_{p \in \mathcal{H}^i(t)} \alpha_{pi}(t) g(\mathbf{u}_p, \mathbf{u}_j) \kappa(t - t_p)$$

$$\underbrace{+ (1 - \beta_{ij}) \sum_{q \in \mathcal{H}^j(t)} \alpha_{qj}(t) g(\mathbf{u}_q, \mathbf{u}_i) \kappa(t - t_q),}_{Neighbor\ Influence}$$

where $g(\cdot)$ is a function measuring the similarity of two nodes, here we define $g(\mathbf{u}_i, \mathbf{u}_j) = -\|\mathbf{u}_i - \mathbf{u}_j\|_2^2$, where other measurements can also be used, such as cosine similarity. $\mathcal{H}^i(t) = \{p\}$ and $\mathcal{H}^j(t) = \{q\}$ are the historical neighbors of node i and j before t, respectively. The term $\kappa(t - t_p) = \exp(-\delta_i(t - t_p))$ is the time decay function with a node-dependent and learnable decay rate $\delta_i > 0$, where t_p is the time of the past event (i, p, t_p). α and β are two attention coefficients determined by a hierarchical temporal attention mechanism, which will be introduced later. A non-linear transfer function $f : \mathbb{R} \to \mathbb{R}_+$ (i.e., exponential function) is designed to ensure a positive value of the intensity, namely,

$$\lambda_{i,j}(t) = f(\tilde{\lambda}_{i,j}(t)). \tag{5.2}$$

The local attention coefficient is defined as follows:

$$\tilde{\alpha}_{pi}(t) = \sigma(\kappa(t - t_p)\mathbf{a}^\top [\mathbf{W}\mathbf{u}_i \oplus \mathbf{W}\mathbf{u}_p]), \tag{5.3}$$

$$\alpha_{pi}(t) = \frac{\exp\left(\tilde{\alpha}_{pi}(t)\right)}{\sum_{p' \in \mathcal{H}^i(t)} \exp\left(\tilde{\alpha}_{p'i}(t)\right)}, \tag{5.4}$$

where $\oplus$ is the concatenation operation. $\mathbf{a} \in \mathbb{R}^{2d}$ serves as the attention vector and $\mathbf{W}$ represents the local weight matrix. Here we incorporate the time decay $\kappa(t - t_p)$ so that if the timestamp t_p is close to t, then node p will have a large impact on the event $o = (i, j, t)$. Similarly, we can get $\alpha_{qj}(t)$ as well.

The global attention of the i's whole neighbors on the current event o as follows:

$$\tilde{\beta}_i = s(\kappa(\overline{t - t_p})\tilde{\mathbf{u}}_i), \quad \tilde{\beta}_j = s(\kappa(\overline{t - t_q})\tilde{\mathbf{u}}_j), \tag{5.5}$$

$$\beta_{ij} = \frac{\exp(\tilde{\beta}_i)}{\exp(\tilde{\beta}_i) + \exp(\tilde{\beta}_j)}, \tag{5.6}$$

where $s(\cdot)$ is a single-layer neural network, which takes the aggregated embedding from neighbors $\tilde{\mathbf{u}}_i$ and the average time decay of past events $\kappa(\overline{t - t_p}) = \exp(-\delta_i(\overline{t - t_p}))$ as input.

Thus, the objective function of micro-dynamics modeling is

$$\mathcal{L}_{mi} = -\sum_{t \in \mathcal{T}} \sum_{(i,j,t) \in \mathcal{E}} \log p(i, j | \mathcal{H}^i(t), \mathcal{H}^j(t)). \tag{5.7}$$

5.2.2.2 Macro-dynamics Preserved Embedding

We define the macro-dynamics which refer to the number of new edges at time t as follows:

$$\Delta e'(t) = n(t)r(t)(\zeta(n(t) - 1)^\gamma), \tag{5.8}$$

where $n(t)$ can be obtained as the network evolves by time t, ζ and γ are learnable with model optimization. As the network evolves by time t, $n(t)$ nodes join in the network. At the next time, each node in the network tries to establish edges with the other $\zeta(n(t) - 1)^\gamma$ nodes with a link rate $r(t)$.

Since the linking rate $r(t)$ plays a vital role in driving the evolution of network scale [112], it is dependent not only on the temporal information but also structural properties of the network. On the one hand, much more edges are built at the inception of the network while the growth rate decays with the densification of the network. Therefore, the linking rate should decay with a temporal term. On the other hand, the establishments of edges promote the evolution of network structures, the linking rate should be associated with the structural properties of the network. Hence, in order to capture such temporal and structural information in network embeddings, we parameterize the linking rate of the network with a temporal fizzling term t^θ and node embeddings:

$$r(t) = \frac{\frac{1}{|\mathcal{E}|} \sum_{(i,j,t) \in \mathcal{E}} \sigma(-\|\mathbf{u}_i - \mathbf{u}_j\|_2^2)}{t^\theta}, \tag{5.9}$$

where θ is the temporal fizzling exponent, $\sigma(x) = \exp(x)/(1 + \exp(x))$ is the sigmoid function.

We learn the parameters in Eq. (5.8) via minimizing the sum of square errors:

$$\mathcal{L}_{ma} = \sum_{t \in \mathcal{T}} (\Delta e(t) - \Delta e'(t))^2, \tag{5.10}$$

where $\Delta e'(t)$ is the predicted number of new edges at time t.

5.2.2.3 The Unified Model

We have the following model to capture the formation process of topological structures and the evolutionary pattern of network scale in a unified manner:

$$\mathcal{L} = \mathcal{L}_{mi} + \epsilon \mathcal{L}_{ma}, \tag{5.11}$$

where $\epsilon \in [0, 1]$ is the weight of the constraint of macro-dynamics on representations learning.

5.2.3 Experiments

5.2.3.1 Experimental Settings

Datasets. We adopt three datasets from different domains, namely, Eucore,[1] DBLP[2], and Tmall.[3] Eucore is generated using email data. The communications between people refer to edges and five departments are treated as labels. DBLP is a co-author network and we take ten research areas as labels. Tmall is extracted from the sales data. We take users and items as nodes, purchases as edges. The five most purchased categories are retained as labels.

Model Settings. We compare the performance of M^2DNE against the following six network embedding methods. including DeepWalk [154], LINE [181], SDNE [191], TNE [255], DynamicTriad (or called Dy.Triad) [254], HTNE [260], and our variant MDNE capturing only micro-dynamics. For a fair comparison, we set the embedding dimension $d = 128$ for all methods. The number of negative samples is set to 5. For M^2DNE, we set the number of historical neighbors to 2, 5, and 2; the balance factor ϵ to 0.3, 0.4, and 0.3 for Eucore, DBLP, and Tmall, respectively.

5.2.3.2 Effectiveness Analysis

Node Classification. After learning the node embeddings on the fully evolved network, we train a logistic regression classifier that takes node embeddings as input features. The ratio of training set is set as 40, 60, and 80%. We report the results in terms of Macro-F1 (i.e., Ma-F1) and Micro-F1 (i.e., Mi-F1) in Table 5.1.

As we can observe, our MDNE and M^2DNE achieve better performance than all baselines in all cases except one. Specifically, compared with methods for static networks (i.e., DeepWalk, LINE, and SDNE), the good performance of MDNE and M^2DNE suggests that the formation process of network structures preserved in our models provides effective information to make the embeddings more discriminative. In terms of methods for temporal networks (i.e., TNE, DynamicTriad, and HTNE), our MDNE and M^2DNE capture the local

---

[1] https://snap.stanford.edu/data/.

[2] https://dblp.uni-trier.de.

[3] https://tianchi.aliyun.com/dataset/.

Table 5.1 Evaluation of node classification. Tr.Ratio means the training ratio

Datasets	Metrics	Ratio (%)	DeepWalk	LINE	SDNE	TNE	Dy.Triad	HTNE	MDNE	M^{2}DNE
Eucore	Ma-F1	40	**0.1878**	0.1765	0.1723	0.0954	0.1486	0.1319	0.1598	0.1365
		60	0.1934	0.1777	0.1834	0.1272	0.1796	0.1731	0.1855	**0.1952**
		80	0.2049	0.1278	0.1987	0.1389	0.1979	0.1927	0.1948	**0.2057**
	Mi-F1	40	0.2089	0.2266	0.2129	0.2298	0.2310	0.2200	0.2273	**0.2311**
		60	0.2245	0.1933	0.2321	0.2377	0.2333	0.2400	0.2501	**0.2533**
		80	0.2400	0.1466	0.2543	0.2432	0.2400	0.2672	0.2702	**0.2800**
DBLP	Ma-F1	40	0.6708	0.6393	0.5225	0.0580	0.6045	0.6768	0.6883	**0.6902**
		60	0.6717	0.6499	0.5498	0.1429	0.6477	0.6824	0.6915	**0.6948**
		80	0.6712	0.6513	0.5998	0.1488	0.6642	0.6836	0.6905	**0.6975**
	Mi-F1	40	0.6653	0.6437	0.5517	0.2872	0.6513	0.6853	0.6892	**0.6923**
		60	0.6689	0.6507	0.5932	0.2931	0.6680	0.6857	0.6922	**0.6947**
		80	0.6638	0.6474	0.6423	0.2951	0.6695	0.6879	0.6924	**0.6971**
Tmall	Ma-F1	40	0.4909	0.4371	0.4845	0.1069	0.4498	0.5481	0.5648	**0.5775**
		60	0.4929	0.4376	0.4989	0.1067	0.4897	0.5489	0.5681	**0.5799**
		80	0.4953	0.4397	0.5312	0.1068	0.5116	0.5493	0.5728	**0.5847**
	Mi-F1	40	0.5711	0.5367	0.5734	0.3647	0.5324	0.6253	0.6344	**0.6421**
		60	0.5734	0.5392	0.5788	0.3638	0.5688	0.6259	0.6369	**0.6438**
		80	0.5778	0.5428	0.5832	0.3642	0.6072	0.6264	0.6401	**0.6465**

and global structures aggregated from neighbors via a hierarchical temporal attention mechanism, which enhances the accuracy of the embeddings of structures. Besides, M^{2}DNE encodes high-level structures in the latent embedding space, which further improves the performance of classification. From a vertical comparison, MDNE and M^{2}DNE continue to perform best against different sizes of training data in almost all cases, which implies the stability and robustness of our models.

Temporal Node Recommendation. For each node v_i in the network before time t, we predict the top-k possible neighbors of v_i at t. We calculate the ranking score as the setting in network reconstruction task, and then derive the top-k nodes with the highest score as candidates. This task is mainly used to evaluate the performance of temporal network embedding methods. However, in order to provide a more comprehensive result, we also compare our method against one popular static method, i.e., DeepWalk.

The experimental results are reported in Fig. 5.3 with respect to Recall@K and Precision@K. We can see that our models MDNE and M^{2}DNE perform better than all the baselines in terms of different metrics. Compared with the best competitors (i.e., HTNE), the recommendation performance of M^{2}DNE improves by 10.88 and 8.34%in terms of Recall@10 and Precision@10 on Eucore. On DBLP, the improvement is 6.05% and 11.69% with respect to Recall@10 and Precision@10. These significant improvements verify that the temporal attention point process proposed in MDNE and M^{2}DNE is capable of modeling fine-grained structures and dynamic patterns of the network. Additionally, the significant

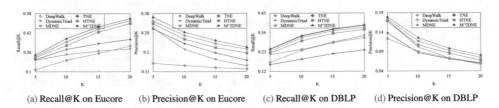

(a) Recall@K on Eucore   (b) Precision@K on Eucore   (c) Recall@K on DBLP   (d) Precision@K on DBLP

Fig. 5.3 Evaluation of temporal node recommendation

improvement of M^{2}DNE benefits from the high-level constraints of macro-dynamics on network embeddings, thus encoding the inherent evolution of the network structure, which is good for temporal prediction tasks.

More detailed description and experimental validation of this work are introduced in [124].

5.3 Heterogeneous Hawkes Process

5.3.1 Overview

Complex systems in real-world scenarios are commonly associated with multiple temporal interactions between different-typed nodes, forming the so-called dynamic heterogeneous graphs. Taking Fig. 5.4b as an example, there are two types of interactions ("co-operation" and "attendance") between two types of nodes (authors and venues) and each interaction is marked with a continuous timestamp to describe when it happened. There have been several heterogeneous graph embedding methods [42, 84, 200, 247], where earlier approaches [42, 54] employ meta-path-based heterogeneous sequences [179] while recent studies [55, 84, 200, 247] apply GNN-based heterogeneous neighborhood aggregation to enhance node representations. However, limited work has been done for embedding dynamic heterogeneous graphs, facing two essential challenges, namely, how to model the continuous dynamics of heterogeneous interactions and how to model the complex influence of different semantics?

Motivated by these challenges, we propose the heterogeneous Hawkes process for dynamic heterogeneous graph embedding (HPGE). To handle the continuous dynamics, we treat heterogeneous interactions as multiple temporal events, which gradually occur over time, and introduce Hawkes process into heterogeneous graph embedding by designing a heterogeneous conditional intensity to model the excitation of historical heterogeneous events to current events. To handle the complex influence of semantics, we further design the heterogeneous evolved attention mechanism which considers both the intra-typed temporal importance of historical events but also the inter-typed temporal impacts from multiple historical events to current type-wise events. Moreover, as current events are influenced more

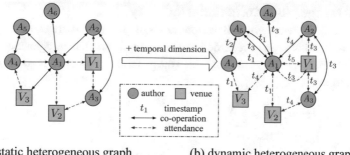

(a) static heterogeneous graph   (b) dynamic heterogeneous graph

Fig. 5.4 Toy examples of static and dynamic heterogeneous graphs

by past important interactions, we adopt the temporal importance sampling strategy to select representative events from historical candidates, balancing their importance and recency.

5.3.2   The HPGE Method

There are three main components of HPGE. First, as shown in Fig. 5.5a, given the respective temporal heterogeneous neighbors of A_1, A_3, and V_1, HPGE evaluates the affinity between each node and its neighbors with a type-wise influence measure. Second, an attentive manner is designed in Fig. 5.5b to capture both the temporal importance of same-typed neighborhoods (intra-att) and the evolution from historical types to the current type (inter-att). For effective and efficient HPGE, we adopt a Temporal Importance Sampling (TIS) strategy in Fig. 5.5c, to extract representative neighbors in both temporal and structural dimensions.

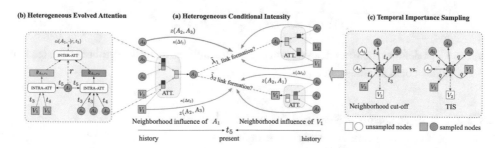

Fig. 5.5 The overall architecture of HPGE

5.3.2.1 Heterogeneous Conditional Intensity

On a dynamic heterogeneous graph, various kinds of interactions are constantly being established over time, which can be regarded as a series of observed heterogeneous events. Given current event $e = (v_i, v_j, t, r)$, we introduce the general heterogeneous conditional intensity function as follows:

$$\tilde{\lambda}(e) = \underbrace{\mu_r(v_i, v_j)}_{\text{base rate}}$$

$$+ \underbrace{\gamma_1 \sum_{r' \in \mathcal{R}} \sum_{p \in N_{i,r',<t}} \alpha(p, e) z(v_p, v_j) \kappa_i(t - t_p)}_{\text{neighborhood influence on source } v_i},$$

$$+ \underbrace{\gamma_2 \sum_{r'' \in \mathcal{R}} \sum_{q \in N_{j,r'',<t}} \alpha(q, e) z(v_q, v_i) \kappa_j(t - t_q)}_{\text{neighborhood influence on target } v_j}$$

(5.12)

where γ_1 and γ_2 are the balance parameters. This conditional intensity function consists of three major parts, including the type-wise base rate, the heterogeneous neighborhood on source node v_i and on target node v_j. At first, given v_i and v_j as well as event type r, the base rate $\mu_r(v_i, v_j)$ is defined as

$$\mu_r(v_i, v_j) = -\sigma(f(\boldsymbol{h}_i \boldsymbol{W}_{\phi(v_i)} - \boldsymbol{h}_j \boldsymbol{W}_{\phi(v_j)}) \boldsymbol{W}_r + b_r),$$

(5.13)

where $\boldsymbol{h}_i \in \mathbb{R}^d$ and $\boldsymbol{h}_j \in \mathbb{R}^d$ are the embedding of v_i and v_j, d is the dimension of node embedding, $\boldsymbol{W}_{\phi(\cdot)} \in \mathbb{R}^{d \times d}$ denotes the type-$\phi(\cdot)$ projection matrix, $f(\cdot)$ denotes the element-level non-negative operation to measure the symmetrical similarity of v_i and v_j, and we adopt self Hadamard product in this paper, namely, $f(\boldsymbol{X}) = \boldsymbol{X} \odot \boldsymbol{X}$, $\boldsymbol{W}_r$ and b_r are the projection and bias of type-r events, $\sigma(\cdot)$ is the non-linear activate function. The relevance between historical neighbors and target nodes are related to their types as well, namely,

$$z(v_p, v_j) = -\|\boldsymbol{h}_p \boldsymbol{W}_{\phi(p)} - \boldsymbol{h}_j \boldsymbol{W}_{\phi(j)}\|_2^2,$$

(5.14)

where $\| \cdot \|_2^2$ denotes the Euclidean distance measure, and the negative symbol indicates that closer nodes could affect greater.

5.3.2.2 Heterogeneous Evolved Attention Mechanism

The importance to current event $\alpha(p, e)$ is defined as

$$\alpha(p, e) = \xi(v_p, t_p | r', v_i, t) \beta(r | r', v_i, t),$$

(5.15)

where r' and r, respectively, denote the type of historical and current event, t_p and t are the corresponding timestamps, $\xi(v_p, t_p | r', v_i, t)$ is the intra-type heterogeneous temporal attention, calculated by

$$\xi(v_p, t_p|r', v_i, t) = \text{softmax}(\sigma(\kappa_i(t-t_p)[h_i W_{\phi(v_i)} \oplus h_j W_{\phi(v_j)}]W_\xi)), \quad (5.16)$$

where $W_\xi \in \mathbb{R}^{2d \times 1}$ denotes the attention projection matrix need to learn, $\oplus$ denotes the concatenation operation, $\text{softmax}(x)$ is in the form of $\exp(x)/\sum_{x'} \exp(x')$. Both the heterogeneity and time decay are taken into consideration. We design the inter-typed $\beta(r|r', v_i, t)$ to model the relevance from historical types to current types, namely,

$$\beta(r|r', v_i, t) = \text{softmax}(\tanh(\tilde{g}_i W_r)w_r)^{\mathsf{T}}, \quad (5.17)$$

where $W_r \in \mathbb{R}^{d|\mathcal{R}| \times d_m}$ and $w_r \in \mathbb{R}^{d_m \times 1}$ are the projection matrices need to learn, d_m is the length of latent dimension and we set $d_m = 0.5d$ here. $\tilde{g}_i$ is the concatenation of historical excitation, namely, $\tilde{g}_i = [\tilde{g}_{i,1} \oplus \tilde{g}_{i,2} \oplus \cdots \oplus \tilde{g}_{i,|\mathcal{R}|}]$, and the sub-excitation from type-r' neighbors is calculated by

$$\tilde{g}_{i,r'} = \sigma\left(\left[\sum_p \xi(v_p, t_p|r', v_i, t)h_p W_{\phi(v_p)}\kappa_i(t-t_p)\right]W_{\beta,r'} + b_{\beta,r'}\right), \quad (5.18)$$

where $W_{\beta,r'} \in \mathbb{R}^{d \times d}$ and $b_{\beta,r'}$ are the projection matrix and bias need to learn. It is naturally a intra-typed attention-based temporal excitation aggregation.

5.3.2.3 Temporal Importance Sampling

As more events are accumulated over time, it becomes expensive to materialize the heterogeneous conditional intensity function. We propose the strategy of Temporal Importance Sampling (TIS) which considers both temporal and structural information to extract representation neighbors. Weighed by the excitation rate and the time decay function, we design the sampler of TIS as follows:

$$q(v_p|v_i, r', t) = \frac{\kappa_i(t-t_p)N_i(v_p)}{\sum_{v_{p'} \in \mathcal{N}_{i,r',<t}} \kappa_i(t-t'_p)N_i(v'_p)}, \quad (5.19)$$

where $q(v_p|v_i, r', t)$ denotes the sampling probability, depending on the importance of node v_h relating to event type r', times of historical occurrence $N_i(v_p)$ as well as time t.

5.3.2.4 Optimization Objective

Given all the historical neighborhoods $\mathcal{N}_{i,<t}$ of v_i and $\mathcal{N}_{j,<t}$ of v_j before time t, the loss of the current event e is defined as follows:

$$\mathcal{L}_{hp}(e) = -\sum_{e \in \mathcal{E}} \log \sigma(\tilde{\lambda}(e)) - \sum_k \mathbb{E}_{j'} \log \sigma(-\tilde{\lambda}(e_{j'})) - \sum_k \mathbb{E}_{i'} \log \sigma(-\tilde{\lambda}(e_{i'})), \quad (5.20)$$

where $e_{i'}$ and $e_{j'}$ are the abbreviations of $e_{i',j,r,t}$ and $e_{i,j',r,t}$, k is the size of negative samples, and $\mathcal{L}_{hp} = \frac{1}{|\mathcal{E}|}\sum_{e \in \mathcal{E}} \mathcal{L}_{hp}(e)$.

Besides, focusing on the downstream tasks like node classification and temporal link prediction, we design the unified loss function as follows:

$$\mathcal{L} = \mathcal{L}_{hp} + \omega_1 \mathcal{L}_{task} + \omega_2 \Omega(\Theta), \tag{5.21}$$

where $\Omega(\Theta)$ is the l2-norm regularization of learnt parameters, $\mathcal{L}_{task}$ is the loss of specific tasks. For node classification and temporal link prediction, we input node embedding or the concatenation of embedding pair into a multi-layer perception to extract the distribution of classifications or the probability of connections.

5.3.3 Experiments

5.3.3.1 Experimental Settings

Datasets and Baselines. We test on three real-world datasets including Aminer, DBLP, and Yelp. We compare the proposed HPGE with three groups of graph embedding models, namely, heterogeneous graph embedding (Metapath2vec [42], HEP [252], HAN [200] and HGT [84]), dynamic graph embedding (CTDNE [139], EvolveGCN [147], and M^{2}DNE [124]), and dynamic heterogeneous graph embedding (DHNE [222], DyHNE [202], and DyHATR [217]).

Parameter Settings. For all methods, we set the embedding dimension $d = 128$, batch size as 1024, learning rate as 0.001, regularization weight $\omega_2 = 0.01$ (if any), and negative sampling size as $k = 5$ (if any). These values give robust performance and are consistent with guidelines from the literature. For HAN, HGT, M^{2}DNE, DyHATR, and our HPGE, we, respectively, limit the size of neighboring candidates to 5, 5, and 10 on the three datasets, using TIS for our method, recency cut-off for M^{2}DNE, and random sampling for others. For dynamic homogeneous baselines, we treat events as homogeneous. For Metapath2Vec and DHNE, we sample sequences via A-A, A-A, and B-U-B on the three datasets, respectively. The other parameters of all baselines follow their original papers. For our HPGE, we set $\gamma_1 = 0.5$ and $\gamma_2 = 0.5$, $\omega_1 = 1$. In addition, the max iteration is set as 500, 500, and 50 on the three datasets.

5.3.3.2 Effectiveness Analysis

Node Classification. This task is to predict the research area of authors on Aminer and DBLP and the category of businesses on Yelp. The train/test ratio is set to 80%/20%. We run all methods five times and evaluate the average Micro-F1 and Macro-F1 scores.

As shown in Table 5.2, our proposed HPGE consistently outperforms all baselines on the three datasets. We make the following observations. (1) Compared with heterogeneous graph embedding approaches (Metapath2vec, HEP, HAN, and HGT), HPGE is able to model the temporal dynamics of heterogeneous events. Similarly, compared to dynamic graph embedding approaches (CTDNE, EvolveGCN, and M^{2}DNE), HPGE benefits from integrating the

Table 5.2 Performance evaluation on node classification. The best performance is bolded and the second best is underlined

Dataset	Aminer		DBLP		Yelp	
Metric	Micro-F1	Macro-F1	Micro-F1	Macro-F1	Micro-F1	Macro-F1
M2V	0.824	0.853	0.874	0.885	0.537	0.642
HEP	0.949	0.952	0.903	0.913	0.622	0.694
HAN	0.967	0.970	0.912	0.914	0.621	0.691
HGT	0.963	0.971	0.920	0.927	0.633	0.705
CTDNE	0.897	0.895	0.872	0.892	0.512	0.639
E.GCN	0.952	0.955	0.887	0.881	0.611	0.687
M2DNE	0.969	0.972	0.891	0.909	0.619	0.693
DHNE	0.901	0.913	0.888	0.909	0.578	0.665
DyHNE	0.970	<u>0.978</u>	0.922	0.922	0.622	<u>0.721</u>
DyHATR	<u>0.973</u>	0.969	<u>0.933</u>	<u>0.935</u>	<u>0.627</u>	0.717
HPGE	**0.988**	**0.984**	**0.951**	**0.952**	**0.649**	**0.731**

abundant semantic information within heterogeneous events. Not surprisingly, the performance gains of HPGE are larger relative to these baselines. (2) Compared with the best competitor DyHATR, which considers both temporal and heterogeneous information, our HPGE can still achieve substantial improvements. The stable improvements demonstrate that modeling the formation process of DHGs can embed evolving nodes better than just paying attention to the evolution between snapshots. (3) Compared with Aminer and DBLP, our model improves more on Yelp. The potential reason is that Yelp is a larger dataset, such that our temporal importance sampling strategy can benefit more.

Temporal Link Prediction. This task is to predict the type-r interaction at time t. Given all temporal heterogeneous events before time t and two nodes v_i and v_j. We treat all events at time t as the positive link, and randomly sample 2 negative instances for both v_i and v_j as the negative links. Subsequently, we test all baselines and our HPGE five times and report the average performance of Accuracy, F1 score, and ROC-AUC in Table 5.3. Obviously, HPGE still achieves the best performance on all datasets. Besides the observations on node classification, HPGE evaluates node proximity based on event types and continuously propagates the influence of types via the temporal point process, while traditional type-wise projections can only model the heterogeneity rather than the interactivity. In addition, HAN, HEP, HGT, DyHNE, DyHATR, and our HPGE always perform better than CTDNE, EvolveGCN, and M^{2}DNE. This phenomenon indicates that integrating semantics into link formation can benefit temporal link prediction more, compared with simply preserving evolving structures.

More detailed descriptions and experimental validation of this work are introduced in [92].

Table 5.3 Performance evaluation on temporal link prediction. The best performance is bolded and the second best is underlined

Dataset	Aminer			Yelp			DBLP		
Metric	ACC	F1	AUC	ACC	F1	AUC	ACC	F1	AUC
M2V	0.806	0.359	0.759	0.790	0.419	0.702	0.798	0.375	0.656
HEP	0.921	0.814	0.944	0.853	0.566	0.829	0.910	0.753	0.934
HAN	0.923	0.811	0.955	0.855	0.591	0.833	0.903	0.751	0.940
HGT	0.938	0.822	0.963	0.859	0.588	0.833	0.899	0.761	0.941
CTDNE	0.824	0.382	0.763	0.806	0.342	0.635	0.713	0.345	0.653
E.GCN	0.904	0.767	0.922	0.822	0.526	0.785	0.853	0.714	0.905
M2DNE	0.929	0.790	0.951	0.854	0.547	0.818	0.896	0.734	0.939
DHNE	0.875	0.634	0.827	0.831	0.504	0.717	0.821	0.668	0.808
DyHNE	0.928	**0.838**	0.959	0.861	0.592	0.831	0.909	0.767	0.940
DyHATR	<u>0.941</u>	0.832	<u>0.966</u>	<u>0.870</u>	<u>0.598</u>	<u>0.843</u>	<u>0.914</u>	<u>0.773</u>	0.936
HPGE	**0.953**	<u>0.835</u>	**0.976**	**0.873**	**0.603**	**0.850**	**0.938**	**0.793**	**0.957**

5.4   Dynamic Meta-Path

5.4.1   Overview

Graphs, such as social networks, e-commerce platforms, and academic graphs, are generally associated with different types and dynamically interact with each other in various ways. As shown in Fig. 5.6a, there are three types of nodes including authors (A), papers (P), and conferences (C) as well as two kinds of interactions, namely, "write" (AP) and "publish" (PC) on an academic graph. Moreover, all timestamps of interactions are recorded as well. These compositions form a typical temporal heterogeneous graph (THG) and contain complex evolution and rich semantics [167, 232]. Focusing on semantic modeling, the recursive neighborhood aggregation in GNNs has been expanded into heterogeneous message passing in recent works [19, 93, 179, 200, 233, 251]. However, these works mostly deal with static structures composited with unvarying relations while neglecting the useful temporal information that exists in most applications. On the other line, dynamic GNNs [125, 127] have been proposed to exploit the temporal information on graphs, where the classical paradigm is to divide the global graph into several independent snapshots [99, 147]. However, such design of dynamic GNNs is not suitable for modeling heterogeneous graphs with dynamic interactions because of overlooking the rich semantics. It is challenging to model the dynamic semantics on THGs and model the mutual evolution of semantics.

In this section, we put forward the dynamic meta-path guided temporal heterogeneous graph neural networks (DyMGNN), to effectively learn node representations on THGs. To

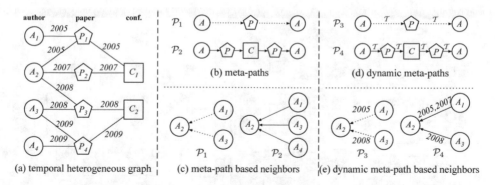

Fig. 5.6 The toy examples of THGs and the comparison of meta-paths and dynamic meta-paths. (a) is the temporal heterogeneous graph consisting of authors, papers and conferences as well as their dynamic interactions. (b) and (c), respectively, showcase the static meta-paths and heterogeneous neighborhoods while (d) and (e) are the corresponding dynamic ones taking temporal bias into consideration

handle the challenge of dynamic semantic modeling, we first propose the dynamic meta-paths to extract the correlations between nodes considering temporal bias, and then propose the dynamic meta-path guided sampling to sample more representative neighbors. To overcome the challenge of semantic-level mutual evolution, we model the evolution of nodes via the heterogeneous mutual evolution attention which captures the evolution of both temporal and semantic levels.

5.4.2   The DyMGNN Method

The overall framework of DyMGNN is shown in Fig. 5.7. Specifically, we first introduce the dynamic meta-path guided temporal importance sampling to sample temporal neighbors for dynamic semantic modeling. As nodes evolve over time and semantic, we then divide THGs into several soft snapshots, which preserve the historical connections to the given snapshots. DyMGNN effectively aggregates semantic-level temporal embedding of nodes (e.g., $h_{i,\mathcal{P},s}$ from path $\mathcal{P}$ at snapshot s) via attention mechanism and temporal encoding. Furthermore, the heterogeneous mutual evolution attention is designed to capture heterogeneous evolving h_i. Finally, the constructed node representation g_i is input into a specific supervised task for optimization.

5.4.2.1   Dynamic Semantic Modeling

Specifically, given a dynamic meta-path $\mathcal{P} : \mathcal{A}_1 \xrightarrow{\mathcal{R}_1,\mathcal{T}} \mathcal{A}_2 \xrightarrow{\mathcal{R}_2,\mathcal{T}} \cdots \xrightarrow{\mathcal{R}_l,\mathcal{T}} \mathcal{A}_{l+1}$, we search the neighbors in reverse and calculate the importance of candidate v_k as follows:

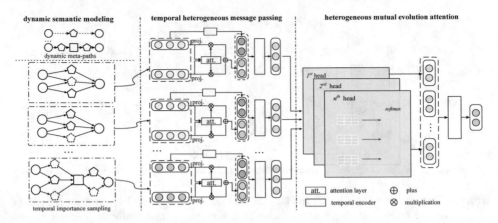

Fig. 5.7 The framework of DyMGNN

$$p(v_k|v_{k+1}, \mathcal{P}) = \begin{cases} 0 & t_k > min_k \\ f(min_k - t_k) & otherwise \end{cases}, \qquad (5.22)$$

where t_k denotes the timestamp of edge $e_{v_k,v_{k+1}}$, $min_k = min(\{t_n|k < n \le l+1, t_n > 0\}$ denotes the minimum timestamp and $f(\cdot)$ is the activation function (e.g., softmax) to evaluate the importance of v_k with timestamp t_k. We further sample dynamic meta-path guided neighbors according to $p(v_k|v_{k+1}, \mathcal{P})$, called temporal importance sampling (TIS).

5.4.2.2 Temporal Heterogeneous Message Passing

Taking the dynamic heterogeneity into consideration, we design the temporal heterogeneous message passing for each dynamic meta-path at each snapshot. Given the neighbors $(v_{j1}, v_{j2}, \ldots, v_{jn})$ of node v_i at a snapshot, we adopt the heterogeneous node-level attention mechanism to enhance or weaken neighborhood information, as follows:

$$a'_{i,j} = \sigma[(\boldsymbol{x}_i \boldsymbol{W}^{ATT}_{\phi(v_i)}||\boldsymbol{x}_j \boldsymbol{W}^{ATT}_{\phi(v_j)}) \boldsymbol{W}^{ATT}_{\psi(v_i,v_j)} + b^{ATT}_{\psi(v_i,v_j)}], \qquad (5.23)$$

where $a'_{i,j} \in \mathbb{R}_+$ is the weight of v_j to v_i, $\boldsymbol{x}_i \in \mathbb{R}^{d_{\phi v_i}}$ is the attribute vector of v_i with dimension $d_{\phi v_i}$, $\boldsymbol{W}^{ATT}_{\phi(v_i)} \in \mathbb{R}^{d_{\phi(v_i)} \times d}$ is the type-wise parameters to project attributes into the latent space, $\boldsymbol{W}^{ATT}_{\psi(v_i,v_j)}$ and $b^{ATT}_{\psi(v_i,v_j)}$ are the latent project matrix and bias of type $\psi(v_i, v_j)$ need to learn. And then, we normalize the attention $a_{i,j,s}$ for meta-path $\mathcal{P}$ at snapshot s as

$$a_{i,j,s} = \frac{a'_{i,j}}{\sum_{v'_j \in Nbr(v_i,s,\mathcal{P})} a'_{i,j'}}, \qquad (5.24)$$

where $Nbr(v_i, s, \mathcal{P})$ is the neighborhood of dynamic meta-path $\mathcal{P}$ at snapshot s. The sub-representation $\boldsymbol{h}'_{i,\mathcal{P},s}$ is

$$h'_{i,\mathcal{P},s} = \text{AGG}(\{a_{i,j,s} \cdot x_j W_{\phi(v_j)} | j \in Nbr(v_i, s, \mathcal{P})\}), \tag{5.25}$$

where $\text{AGG}(\cdot)$ is the pooling function and we select *sum* pooling in this paper. Notice that, as the attributes of different-typed nodes belong to different spaces, we adopt $W_{\phi(x_j)}$ to project all attributes in the same latent space, which is labeled "proj" in Fig. 5.7. Inspired by [214], we construct the embedding of v_i at sth snapshot of meta-path $\mathcal{P}$ with a temporal encoder as

$$h_{i,\mathcal{P},s} = \sigma((h'_{i,\mathcal{P},s} + \mathcal{K}(T - t_s)) W_S + b_S), \tag{5.26}$$

where $W_S \in \mathbb{R}^{d \times d}$ and $b_S \in \mathbb{R}$ are learnable parameters, and $\mathcal{K}(\cdot)$ is the temporal encoder,

5.4.2.3 Heterogeneous Mutual Evolution Attention

As is mentioned, given different dynamic meta-paths, there are multiple semantics-level representations of nodes on THGs (e.g., preferences or interests) which mutually evolve over time. To model the evolution of preferences, we propose the heterogeneous mutual evolution attention to detect the potential dependence of mutual evolution of nodes at semantic level. Given the embedding matrix h_i of node v_i with dimension $N_S \times N_{\mathcal{P}} \times d$ where N_S is the number of snapshots, we, respectively, calculate $Q_i = h_i W_Q$, $K_i = h_i W_K$, and $V_i = h_i W_V$. And then, All Q_i, K_i, and V_i are divided into N_h heads, and the attention $att_{i,n}$ of node v_i of head n is

$$att_{i,n} = softmax(Q_{i,n}^T K_{i,n} / \sqrt{d/n}), \tag{5.27}$$

and then, we concatenate all sub-embeddings h_i as follows:

$$h_i = ||_{n=0}^{N_h} (att_{i,n} \cdot V_{i,n}). \tag{5.28}$$

Finally, the node representation is defined as

$$g_i = [h_i || x_i] W_{\phi(v_i),G} + b_{\phi(v_i),G}, \tag{5.29}$$

where $W_{\phi(v_i),G}$ and $b_{\phi(v_i),G}$ are the learnable type-aware project matrix and bias. Different from self-attention mechanisms [230], the mutual evolution attention takes into account both the dynamics and semantics besides inherent attributes.

5.4.2.4 Optimization Objective

The overall cross-entropy loss is defined as

$$\mathcal{L} = \sum_{i,z} -y_{i,z} log(\hat{y}_{i,z}) - (1 - y_{i,z}) log(1 - \hat{y}_{i,z}) + \alpha \Omega(\Theta), \tag{5.30}$$

where y is the ground truth, $\hat{y}$ is the prediction, $\Omega(\Theta)$ denotes the regularization of total parameters Θ to avoid over-fitting and α is the rate. For link prediction, i and z are two nodes and we generate the probability $MLP(g_i \| g_z)$ of the connections between two nodes. Notice that we adopt Adam [101] to minimize Equation (5.30).

5.4.3   Experiments

Datasets and Baselines. We test our model on three real-world THGs including Aminer academic graph, DBLP academic graph, and Yelp business graph. There are three types of eight representative baselines, including dynamic GNNs (DGNN [125], EvolvGCN [147], M^2DNE [124] and TGAT [214]), heterogeneous GNNs (HAN [200] and HGT [84]), and dynamic heterogeneous graph approaches (DyHNE [202] and DyHATR [217]).

Parameter Settings. For Aminer, DBLP, and Yelp datasets, we set the time span of each snapshot as 1 year, the number of dynamic meta-path-based neighbors as 5 for all three datasets. We, respectively, set the number of snapshots N_S as 10, 6, and 10. For all the baselines and DyMGNN, we set the max iteration as 200, the dimension of nodes $d = 128$, the learning rate as 0.001, the weight of regularization $\alpha = 0.001$. The size of each mini-batch is set as 128. The remainder parameters of baselines are set following the original papers. For all homogeneous graph models (DGNN, EvolvGCN, M^2DNE and TGAT), we remove the types of edges.

5.4.3.1 Effectiveness Analysis

Node Classification. In this task, we, respectively, consider the research domains of authors on Aminer and DBLP datasets and the categories of businesses on Yelp dataset as labels. We train the model with training instances of different scales (i.e., 40, 60, and 80%). We report the results in terms of Micro-F1 (Mi-F1) and Macro-F1 (Ma-F1) in Table 5.4 and make the following observations.

First, DyMGNN consistently achieves the best performance on all three datasets with different-scale training instances. Compared with dynamic homogeneous approaches (i.e., DGNN, E.GCN, TGAT, and M^2DNE), the advantage is in naturally utilizing semantics rather than single-typed edges. Compared with heterogeneous models including static HAN and HGT as well as dynamic DyHNE and DyHATR, DyMGNN has the ability to model fine-grained mutual evolution between different semantics. Second, semantic attention mechanism illustrates the advantages. While DyHNE treats all semantics as the equally important, the semantic attention-based DyHATR and our DyMGNN usually achieve better performance due to their ability to evaluate the importance of different semantics.

Temporal Link Prediction. In this task, we focus on predicting the "APA" links on both Aminer and DBLP dataset and the "BRURB" links on Yelp dataset. We set the interactions in the recent year (for Aminer and DBLP) or the last quarter (for Yelp). As can be seen in

Table 5.4 The performance of methods for node classification. The best performance is in bold and the second best is underlined

Dataset	Metric	Ratio (%)	Methods								
			DGNN	E.GCN	M²DNE	TGAT	HAN	HGT	DyHNE	DyHATR	DyMGNN
Aminer	Mi-F1	40	0.779	0.819	0.826	0.835	0.868	0.872	<u>0.884</u>	0.877	**0.925**
		60	0.795	0.835	0.830	0.841	0.880	0.892	0.895	<u>0.899</u>	**0.925**
		80	0.812	0.861	0.834	0.850	0.901	0.906	<u>0.918</u>	0.907	**0.947**
	Ma-F1	40	0.794	0.814	0.811	0.829	0.855	0.865	<u>0.876</u>	0.872	**0.923**
		60	0.817	0.821	0.828	0.840	0.871	0.889	0.897	<u>0.902</u>	**0.922**
		80	0.834	0.845	0.829	0.846	0.892	0.918	0.913	<u>0.932</u>	**0.944**
DBLP	Mi-F1	40	0.589	0.659	0.686	0.677	0.698	0.693	0.690	<u>0.700</u>	**0.717**
		60	0.623	0.672	0.701	0.680	0.712	0.717	0.702	<u>0.728</u>	**0.739**
		80	0.644	0.679	0.710	0.691	0.724	0.720	<u>0.733</u>	0.726	**0.745**
	Ma-F1	40	0.581	0.632	0.657	0.649	0.666	0.658	0.642	<u>0.671</u>	**0.689**
		60	0.633	0.658	0.670	0.651	0.684	0.686	0.654	<u>0.689</u>	**0.705**
		80	0.652	0.666	0.688	0.670	0.691	0.694	0.692	<u>0.697</u>	**0.711**
Yelp	Mi-F1	40	0.566	0.592	0.601	0.585	0.620	0.628	0.616	<u>0.633</u>	**0.651**
		60	0.572	0.607	0.602	0.590	0.631	<u>0.658</u>	0.625	0.638	**0.672**
		80	0.587	0.619	0.610	0.608	0.644	0.648	<u>0.652</u>	0.641	**0.662**
	Ma-F1	40	0.540	0.577	0.569	0.545	0.600	<u>0.609</u>	0.607	0.609	**0.621**
		60	0.555	0.582	0.570	0.570	0.610	<u>0.616</u>	0.615	0.612	**0.628**
		80	0.563	0.590	0.579	0.572	0.618	0.632	0.629	<u>0.633</u>	**0.643**

Table 5.5 The performance of methods for temporal link prediction. The best performance is in bold and the second best is underlined. Relative improvements of DyMGNN w.r.t. the second best are reported as well

Dataset	Metric	Methods									Improv. (%)
		DGNN	E.GCN	M²DNE	TGAT	HAN	HGT	DyHNE	DyHATR	DyMGNN	
Aminer	F1	0.744	0.747	0.750	0.772	0.764	0.772	0.789	0.785	**0.799**	1.3
	PR-AUC	0.769	0.766	0.778	0.800	0.795	0.803	0.815	0.809	**0.820**	0.6
	ROC-AUC	0.838	0.782	0.848	0.880	0.877	0.882	0.893	0.882	**0.900**	0.8
DBLP	F1	0.610	0.625	0.606	0.616	0.634	0.639	0.642	0.655	**0.662**	1.1
	PR-AUC	0.629	0.638	0.636	0.648	0.656	0.652	0.654	0.663	**0.673**	1.5
	ROC-AUC	0.664	0.669	0.679	0.684	0.683	0.681	0.685	0.690	**0.706**	2.3
Yelp	F1	0.579	0.618	0.594	0.599	0.605	0.610	0.616	0.626	**0.643**	2.7
	PR-AUC	0.616	0.629	0.628	0.613	0.647	0.652	0.648	0.657	**0.676**	2.9
	ROC-AUC	0.635	0.658	0.654	0.647	0.669	0.672	0.664	0.670	**0.685**	1.9

Table 5.5, we can find that DyMGNN performs consistently better than all baselines, and the improvement rate to the second one is from 0.8 to 2.3% in ROC-AUC metric. Moreover, the dynamic heterogeneous alternatives DyHNE and DyHATR both perform better than other baselines on the three datasets, which verify the superiority of modeling temporal and heterogeneous information again. In addition, DyHNE outperforms DyHATR on Aminer dataset while performing worse on DBLP and Yelp datasets. However, DyMGNN keeps the advantages on all three datasets, indicating the stability of our model.

More detailed description and experimental validation of this work are introduced in [91].

5.5   Conclusion

In this section, we have introduced the GNN-based dynamic graph representation learning methods. We make the first attempt to explore temporal network embedding from microscopic and macroscopic perspectives by designing a temporal attention point process to capture structural and temporal properties at a fine-grained level. And then, we propose the HPGE model which introduces Hawkes process to effectively and efficiently model the formation process of temporal heterogeneous events via heterogeneous conditional intensity function and the temporal importance sampling strategy. Furthermore, we design a novel graph neural network model called DyMGNN to handle dynamic semantic and semantic-level mutual evolution modeling via dynamic meta-path-based temporal importance sampling and the heterogeneous mutual evolution mechanism. Experimental results on real-world public datasets demonstrate that our proposed methods consistently outperform state-of-the-art baselines.

5.6   Further Reading

The research of dynamic graph modeling is gradually emerging currently and has been widely used in many scenarios (e.g., recommender systems [237] and anomaly detection [16]). The most recent work can be roughly divided into two major categories, including homogeneous dynamic graph modeling [44, 152, 178] and heterogeneous dynamic graph modeling [85, 253].

For homogeneous dynamic graph modeling, [152] introduces reinforcement learning into traditional snapshot-RNN-based dynamic GNNs, aiming at handling data sparsity of incomplete dynamic graphs. Considering that traditional representation learning techniques would suffer from the rapidly increased scale of dynamic graphs, [178] proposes to learn dynamic graph representation in hyperbolic space so as to bridge this gap. Furthermore, [44] pays attention to the complex mutual impacts of temporal and structural factors of dynamic graphs and proposes the spatial-temporal GNN to extract the latent interplay among them for keeping higher interpretability. For heterogeneous dynamic graph modeling, [253] models the heterogeneous dynamics by combining with the residual compressed aggregation component. Paying attention to the types of nodes, [85] learns multi-type node embeddings by integrating the Hawkes process and hierarchical attention mechanism.

In addition, many advanced techniques of dynamic and heterogeneous modeling are summarized in [14, 99, 197].

Hyperbolic Graph Neural Networks

6

Yiding Zhang

6.1 Introduction

The real-world data usually comes together with the graph structure, such as social networks, citation networks, biology networks. Graph neural network (GNN) [65, 162], as a powerful deep representation learning method for such graph data, has shown superior performance on network analysis and aroused considerable research interest. There have been many studies using neural networks to handle the graph data. For example, [65, 162] leveraged deep neural network to learn node representations based on node features and the graph structure; [37, 74, 102] proposed the graph convolutional networks by generalizing the convolutional operation to graph; [189] designed a novel convolution-style graph neural network by employing the attention mechanism in GNNs. These proposed GNNs have been widely used to solve many real-world application problems. Ying et al. [223], Song et al. [173] and Fan et al. [46] proposed recommender systems based on GNNs. Parisot et al. [148] and Rhee et al. [158] leveraged graph convolution to solve disease prediction problem.

Essentially, most of the existing GNN models are primarily designed for the graphs in Euclidean spaces. The main reason is that Euclidean space is the natural generalization of our intuition-friendly and visible three-dimensional space. However, real-world spatial structured data can be non-Euclidean surfaces (e.g., hyperbolic spaces) [11, 206]. For example, biologists may inspect the geometric shape of a protein surface to determine its interaction with other biomolecules for drug discovery; physicists find some statistical mechanics (e.g., heterogeneous degree distributions) of complex networks can be naturally discovered in hyperbolic spaces [105]; In such cases, existing models that assume spatial structure to be

Y. Zhang (✉)
Beijing University of Posts and Telecommunications, Beijing, China

on a Euclidean plane may fail to achieve satisfied performance. On the other hand, several works [105, 141] have demonstrated that the hyperbolic spaces could be the latent spaces of graph data, as the hyperbolic space may reflect some properties of graph naturally, e.g., hierarchical and scale-free structure [105, 134]. Inspired by this insight, the study of graph data in hyperbolic spaces has received increasing attentions, such as hyperbolic graph embedding [58, 141, 142, 161].

In this chapter, we will introduce three hyperbolic GNNs. Firstly, hyperbolic graph attention network (named HAT) is designed to learn hyperbolic graph representations in hyperbolic spaces based on attention mechanism. Secondly, to learn better hyperbolic graph representations, Lorentzian graph convolutional network (named LGCN) is proposed, which rigorously guarantees the learned node features follow the hyperbolic geometries. Thirdly, hyperbolic heterogeneous network embedding (named HHNE) is proposed to preserve the structure and semantic information of HG in hyperbolic spaces.

6.2   Hyperbolic Graph Attention Network

6.2.1   Overview

Hyperbolic graph representation learning has shown very promising results [58, 141, 142, 161]. However, there are two key challenges in designing the graph attention network in hyperbolic spaces: (1) One is that there are many different procedures in GNNs, e.g., the projection step, the attention mechanism, and the propagation step. However, different from Euclidean spaces, hyperbolic spaces are not vector spaces, so the vector operations (e.g., including vector addition, subtraction, and scalar multiplication) cannot be carried in hyperbolic spaces. Although there are some attempts in designing some hyperbolic graph operations, e.g., feature transformation [21, 120], it is unclear how to design the multi-head attention mechanism in the hyperbolic setting, which is a key step in GAT (graph attention network) [189]. Also, as hyperbolic spaces have negative curvatures, choosing a proper curvature is needed for our model. How can we effectively implement those hyperbolic graph operations of GNN, especially for multi-head attention, in an elegant way? (2) Another challenge is that mathematical operations in hyperbolic spaces could be more complex than those in Euclidean spaces. Some basic properties of mathematical operations, such as the commutative or associative of "vector addition", are not satisfied anymore in hyperbolic spaces. How can we assure the learning efficiency in the proposed model?

In this section, we exploit the graph attention network to learn node representations of graphs in hyperbolic spaces. As the gyrovector space framework provides an elegant algebraic formalism for hyperbolic geometry, we utilize this framework to learn the graph representations in hyperbolic spaces. Specifically, we first use the operations defined in the framework to transform the features in a graph; and we exploit the proximity in the product of hyperbolic spaces to model the multi-head attention mechanism in the non-Euclidean setting; afterward, we further devise a parallel strategy using logarithmic and exponential maps to improve the efficiency of our proposed model. The comprehensive experimental results demonstrate the effectiveness of the proposed model, compared with state-of-the-art methods.

6.2.2   The HAT Method

6.2.2.1 The HAT Framework

In this section, we present our hyperbolic graph attention network model, named HAT, whose framework is shown in Fig. 6.1. Hence, our model can be summarized as two procedures: (1) The hyperbolic feature projection. Given the original input node feature, this procedure projects it into a hyperbolic space through the exponential map and the hyperbolic linear transformation, so as to obtain the latent representation of the node in hyperbolic space. (2) The hyperbolic attention mechanism. This procedure designs an attention mechanism based on the hyperbolic proximity to aggregate the latent representations. Also, we devise an acceleration strategy to speed up the proposed model by using logarithmic and exponential map, since operations in hyperbolic spaces are usually more complex and time consuming than that in Euclidean spaces. Moreover, we leverage the product of hyperbolic spaces

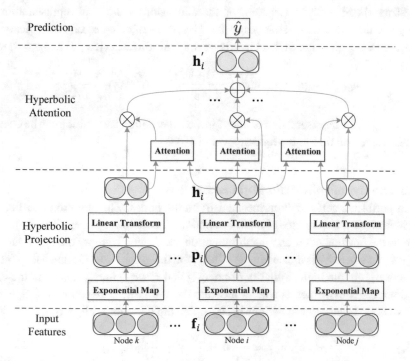

Fig. 6.1 The framework of HAT model

to achieve the multi-head attention. Finally, we feed the aggregated representations to a loss function for the downstream task. Here we mainly describe a single graph attentional layer, as the sole layer is used throughout all of our proposed HAT architectures in our experiments. Also, all the graph operations are built upon an open d-dimensional Poincaré ball: $\mathbb{D}_c^d := \{\mathbf{x} \in \mathbb{R}^d : c\|\mathbf{x}\|^2 < 1\}$, with the Riemannian metric: $g_\mathbf{x} = (\lambda_\mathbf{x}^c)^2 g^\mathbb{R}$.

6.2.2.2 The Hyperbolic Feature Projection

The input of GNN is the node feature, which usually lives in Euclidean spaces. To make the node feature available in hyperbolic spaces, we use the exponential map to project the feature into the hyperbolic spaces [21, 120].

Specifically, let $\mathbf{f}_i$ be the feature of node i, and then for $\mathbf{f}_i \in T_\mathbf{x}\mathbb{D}_c^d \backslash \{\mathbf{0}\}$, where $\mathbf{x}$ is a point in hyperbolic spaces and $T_\mathbf{x}\mathbb{D}_c^d$ is the tangent space at point $\mathbf{x}$, when $\mathbf{x} = \mathbf{0}$, the exponential map is defined as

$$\exp_\mathbf{0}^c(\mathbf{f}_i) = \tanh(\sqrt{c}\|\mathbf{f}_i\|)\frac{\mathbf{f}_i}{\sqrt{c}\|\mathbf{f}_i\|}. \tag{6.1}$$

Here we assume that the feature $\mathbf{f}_i$ lies in the tangent spaces at the point $\mathbf{x} = \mathbf{0}$, so we can get the new feature $\mathbf{p}_i \in \mathbb{D}_c^d$ in hyperbolic spaces via $\mathbf{p}_i = \exp_\mathbf{0}^c(\mathbf{f}_i)$.

We then transform $\mathbf{p}_i$ into a higher level latent representation $\mathbf{h}_i$ to obtain sufficient representation power. To this end, we use a shared linear transformation parametrized by a weight matrix $\mathbf{M} \in \mathbb{R}^{d' \times d}$ (where d' is the dimension of the final representation). The challenge is that we cannot simply use the Euclidean matrix-vector multiplication, and instead, here we employ the Möbius matrix-vector multiplication [59]. If $\mathbf{Mp}_i \neq \mathbf{0}$, we have

$$\mathbf{h}_i = \mathbf{M} \otimes_c \mathbf{p}_i = \frac{1}{\sqrt{c}} \tanh\left(\frac{\|\mathbf{Mp}_i\|}{\|\mathbf{p}_i\|} \tanh^{-1}(\sqrt{c}\|\mathbf{p}_i\|)\right) \frac{\mathbf{Mp}_i}{\|\mathbf{Mp}_i\|}. \tag{6.2}$$

Here $\mathbf{h}_i \in \mathbb{D}_c^{d'}$ is the representation of node i in the hyperbolic space, which can be considered as a latent representation in the hidden layer of HAT.

6.2.2.3 The Hyperbolic Attention Mechanism

We then perform a self-attention mechanism on the nodes. The attention coefficient α_{ij}, which indicates the importance of node j to node i, is as follows: $\alpha_{ij} = f(\mathbf{h}_i, \mathbf{h}_j)$, where f represents the function of computing the attention coefficient. Here we only compute α_{ij} for nodes $j \in N_i$, where N_i is the neighbors of node i in the graph. Considering a large attention coefficient α_{ij} for the high similarity of nodes j and i, we define f based on the distance in hyperbolic spaces, which is able to measure the similarity between nodes. Specifically, given two node latent representations $\mathbf{h}_i, \mathbf{h}_j \in \mathbb{D}_c^d$, the distance is given by:

$$d_c(\mathbf{h}_i, \mathbf{h}_j) = \frac{2}{\sqrt{c}} \tanh^{-1}(\sqrt{c}\| - \mathbf{h}_i \oplus_c \mathbf{h}_j\|), \tag{6.3}$$

where the operator $\oplus_c$ is the Möbius addition in $\mathbb{D}_c^d$ as $\mathbf{h}_i \oplus_c \mathbf{h}_j :=$ $\frac{(1+2c\langle \mathbf{h}_i, \mathbf{h}_j \rangle + c\|\mathbf{h}_j\|^2)\mathbf{h}_i + (1-c\|\mathbf{h}_i\|^2)\mathbf{h}_j}{1+2c\langle \mathbf{h}_i, \mathbf{h}_j \rangle + c^2\|\mathbf{h}_i\|^2\|\mathbf{h}_j\|^2}$. Then, we perform the self-attention coefficient as $\alpha_{ij} =$ $-d_c(\mathbf{h}_i, \mathbf{h}_j)$. For all the neighbors of node i (including itself), we should make their attention coefficients easily comparable, so we normalize them using the softmax function: $w_{ij} = \frac{\exp(\alpha_{ij})}{\sum_{l \in N_i} \exp(\alpha_{il})}$. The normalized attention coefficient w_{ij} is used to compute a linear combination of the latent representations of all the nodes $j \in N_i$. So the final aggregated representation $\mathbf{h}_i'$ for node i is as follow:

$$\mathbf{h}_i' = \sigma^{\otimes_c}\left(\sum_{j \in N_i}^{\oplus_c} w_{ij} \otimes_c \mathbf{h}_j\right), \tag{6.4}$$

where the $\sum^{\oplus_c}$ is the accumulation of Möbius addition. $\sigma^{\otimes_c}$ is hyperbolic non-linearity. The operation $w_{ij} \otimes_c \mathbf{h}_j$ can be realized by the Möbius scalar multiplication. The Möbius scalar multiplication of $\mathbf{h}_j \in \mathbb{D}_c^d \setminus \{0\}$ by $w_{ij} \in \mathbb{R}$ is defined as $w_{ij} \otimes_c \mathbf{h}_j :=$ $\frac{1}{\sqrt{c}} \tanh\left(w_{ij} \tanh^{-1}(\sqrt{c}\|\mathbf{h}_j\|)\right) \frac{\mathbf{h}_j}{\|\mathbf{h}_j\|}$.

6.2.2.4 Multi-head Attention

Multi-head attention can make model get better results [189]. We aim to extend our self-attention mechanism to multi-head attention. However, it is unclear how to design the multi-head attention in hyperbolic spaces. The main challenge of designing hyperbolic multi-head attention is to realize the concatenation operation in an elegant way. To bridge this gap, we should design the multi-head attention in the product of hyperbolic spaces. Specifically, for HAT with K-head attention, some operations, including input feature, hyperbolic projection (as shown in Fig. 6.1), are carried independently for each attention without any change. For the hyperbolic attention part, as the node feature lives in the product of K m-dimensional hyperbolic spaces $\mathbb{D}^m \times \mathbb{D}^m \cdots$, we denote the latent representation of node i as H_i, which is composed of K m-dimensional features $\mathbf{h}_i^{(1)}, \mathbf{h}_i^{(2)}, \ldots, \mathbf{h}_i^{(K)}$: $H_i = \big\|_{k=1}^K \mathbf{h}_i^{(k)}$. The attention coefficient of hyperbolic multi-head attention α_{ij}' can be computed as $\alpha_{ij}' = -d_p(H_i, H_j)$, where $d_p(\cdot, \cdot)$ is the distance in the product of hyperbolic spaces. Given node latent representations H_i, H_j, this distance is defined as $d_p(H_i, H_j) = \sqrt{\sum_{k=1}^K d_c^2(\mathbf{h}_i^{(k)}, \mathbf{h}_j^{(k)})}$, which is also a metric distance. The proof is shown in the appendix. Also, we normalize the attention coefficients for all the neighbors of node i via softmax function: $w_{ij}' = \frac{\exp(\alpha_{ij}')}{\sum_{l \in N_i} \exp(\alpha_{il}')}$. Next, we aggregate the node representation via the acceleration strategy for each attention independently:

$$\mathbf{h}_i'^{(k)} = \sigma^{\otimes_c}\left(\sum_{j \in N_i}^{\oplus_c} w_{ij}' \otimes_c \mathbf{h}_j'^{(k)}\right). \tag{6.5}$$

Thus, the final output of HAT with multi-head attention is given as $H_i' = \big\|_{k=1}^K \mathbf{h'}_i^{(k)}$.

6.2.2.5 Acceleration of HAT

In our proposed model HAT, the calculation of Eqs. (6.4) and (6.5) is very time consuming, which seriously affects the efficiency of HAT. As mentioned before, the Möbius addition in Eqs. (6.4) and (6.5) is neither commutative nor associative, meaning that we have to calculate the results by order. Specifically, taking Eq. (6.4) as an example, we denote $w_{ij} \otimes_c \mathbf{h}_j$ as $\mathbf{v}_{ij}$, so the accumulation term in Eq. (6.4) can be rewritten as follows:

$$\sum_{j \in N_i}^{\oplus_c} \mathbf{v}_{ij} = \mathbf{v}_{i1} \oplus_c \mathbf{v}_{i2} \oplus_c \mathbf{v}_{i3} \oplus_c \cdots = \Big(\big((\mathbf{v}_{i1} \oplus_c \mathbf{v}_{i2}) \oplus_c \mathbf{v}_{i3} \big) \oplus_c \cdots \Big). \tag{6.6}$$

As we can see, the calculation of Eq. (6.6) has to be in a serial manner. It is well known that there are always some hubs that have many edges in a large graph, so the calculation becomes very impractical.

The Möbius version of operation [59] provides a feasible way to solve this problem. Actually, some operations in gyrovector spaces can be derived with logarithmic map and exponential map [59]. Taking the Möbius scalar multiplication as an example, it first uses the logarithmic map to project the representation into a tangent space, and then multiply the projected representation by a scalar in the tangent space, and finally project it back on the manifold with the exponential map [59]. The logarithmic and exponential map can project the node representations between the two manifolds in a correct manner. Specifically, for two points $\mathbf{v}_i \in \mathbb{D}_c^d$ and $\mathbf{v}_j \in \mathbb{D}_c^d \backslash \{\mathbf{0}\}$, the logarithmic map $\log_{v_i}^c : \mathbb{D}_c^n \to T_{v_i} \mathbb{D}_c^n$ is given for $\mathbf{v}_i = \mathbf{0}$ by: $\log_{\mathbf{0}}^c(\mathbf{v}_j) = \tanh^{-1}(\sqrt{c}\|\mathbf{v}_j\|) \frac{\mathbf{v}_j}{\|\mathbf{v}_j\|}$. The logarithmic map enables us to get the representation $\log_{\mathbf{0}}^c(\mathbf{v}_j)$ in a tangent space. As the tangent spaces are vector spaces, we can combine the representations, just as we do it in the Euclidean spaces, $\sum_{j \in N_i} \log_{\mathbf{0}}^c (w_{ij} \otimes_c \mathbf{h}_j)$. After the linear combination, we use the exponential map to project the representations back to the hyperbolic spaces, giving rise to the final representation as [21, 120]:

$$\mathbf{h}_i' = \sigma^{\otimes_c} \Big(\exp_{\mathbf{0}}^c \Big(\sum_{j \in N_i} \log_{\mathbf{0}}^c \big(w_{ij} \otimes_c \mathbf{h}_j \big) \Big) \Big). \tag{6.7}$$

The diagram of this operation is shown in Fig. 6.2. Different from Eqs. (6.4) and (6.5), the accumulation operation in the Eq. (6.7) is commutative and associative, so it can be computed in a parallel way. Thus, our model becomes more efficient. Moreover, following [21, 105], we can also change the curvature of hyperbolic spaces to make HAT learn the curvature of hyperbolic spaces for a given graph.

6.2.3 Experiments

6.2.3.1 Experimental Settings

Datasets. We employ four widely used real-world graphs for evaluations, including Cora, Citeseer, Pubmed [163] and Amazon Photo [165]. To quantify which datasets are suitable to be modeled in hyperbolic spaces. We also compute δ_{avg}-hyperbolicity [69] to quantify

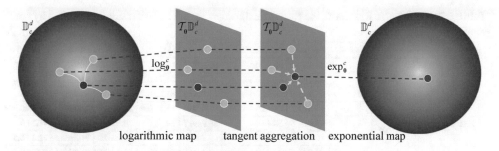

logarithmic map   tangent aggregation   exponential map

Fig. 6.2 The acceleration of HAT via tangent aggregation. The node representations are mapped between hyperbolic and Euclidean spaces by using logarithmic and exponential map, so the representations of neighbors can be aggregated in the tangent space

the tree-likeliness of these datasets. A low δ_{avg}-hyperbolicity of a graph indicates that it has an underlying hyperbolic geometry (Table 6.1).

Baselines. We compare our method with the following state-of-the-art methods: DeepWalk [154], PoincaréEmb [141], GCN [102], GAT [189], HGNN [120], HGCN [21].

6.2.3.2 Node Classification

Node classification is a basic task widely used to evaluate the embedding effectiveness. GCN, GAT, and HAT are the semi-supervised models which can be directly used to classify the nodes. For DeepWalk, we employ KNN classifier with $k = 5$ to perform the node classification. We report the average Micro-F1 of 10 runs with random weight initialization.

The results are shown in Table 6.2. It is obvious that HAT achieves the best performance in most cases, and its superiority is more significant for the low dimension setting. Specifically, as we can infer from Table 6.2, compared with the second best results, HAT achieves better results on these 4 datasets with 8 dimensions. Also, we compute δ_{avg} for the 4 datasets. A lower δ_{avg}-hyperbolicity of a graph indicates that this graph is more suitable to be embedded in the hyperbolic spaces. The performance of the hyperbolic GNNs (i.e., HGNN, HGCN, and HAT) roughly meet this point. Compared with the Euclidean GNNs (i.e., GCN and GAT), the hyperbolic GNNs perform better in Amazon Photo, that because Amazon Photo has a low

Table 6.1 Summary of the datasets

Dataset	Cora	Citeseer	Pubmed	Amazon photo
# Nodes	2708	3327	19717	7650
# Edges	5429	4732	44338	143663
# Features	1433	3703	500	745
# Classes	7	6	3	8

Table 6.2 Quantitative Results on the Node Classification Task. The best results are marked by bold numbers

Dataset	dimension	Deepwalk	PoincaréEmb	GCN	GAT	HGNN	HGCN	HAT
Cora $\delta_{avg} = 0.353$	8	64.5±1.2	57.5±0.6	80.3±0.8	80.4±0.8	80.4±1.2	80.0±0.7	**82.6±0.7**
	16	65.2±1.6	64.4±0.3	81.9±0.6	81.7±0.7	81.6±0.8	81.3±0.6	**83.3±0.6**
	32	65.9±1.5	64.9±0.4	81.5±0.4	82.6±0.7	81.3±0.6	81.7±0.7	**83.6±0.5**
	64	66.5±1.7	68.6±0.4	81.6±0.4	83.1±0.6	81.9±0.7	81.4±0.6	**83.4±0.5**
Citeseer $\delta_{avg} = 0.461$	8	47.8±1.6	38.6±0.4	68.9±0.7	69.5±0.8	70.6±0.9	70.9±0.6	**71.3±0.7**
	16	46.2±1.5	40.4±0.5	69.8±0.5	70.4±0.7	71.0±0.8	71.2±0.5	**72.2±0.6**
	32	43.6±1.9	43.5±0.5	70.4±0.5	71.9±0.7	71.8±0.5	71.9±0.4	**72.2±0.4**
	64	46.6±1.4	43.6±0.4	70.8±0.4	**72.4±0.7**	71.5±0.5	71.7±0.5	72.1±0.4
Pubmed $\delta_{avg} = 0.355$	8	73.2±0.7	66.0±0.8	78.6±0.4	71.9±0.7	75.6±0.4	77.9±0.6	**78.9±0.8**
	16	73.9±0.8	68.0±0.4	79.1±0.5	75.9±0.7	78.3±0.6	78.4±0.4	**79.3±0.5**
	32	72.4±1.0	68.4±0.5	78.7±0.5	78.2±0.6	78.7±0.4	78.6±0.6	**79.6±0.5**
	64	73.5±1.0	69.9±0.6	79.1±0.5	78.7±0.4	78.7±0.5	79.3±0.5	**79.5±0.5**
Amazon Photo $\delta_{avg} = 0.268$	8	76.2±0.7	77.9±0.9	84.1±0.9	81.9±0.9	84.2±1.3	86.7±0.8	**88.7±0.5**
	16	78.9±0.8	78.6±0.9	86.0±0.7	83.4±0.8	85.9±0.8	86.3±0.5	**89.3±0.5**
	32	81.7±1.0	76.2±0.8	86.7±0.7	84.3±0.7	86.5±0.5	87.9±0.4	**89.7±0.5**
	64	77.5±1.0	78.6±0.6	86.9±0.6	84.5±0.7	87.9±0.8	88.9±0.4	**89.4±0.4**

δ_{avg}-hyperbolicity. One the other hand, because Citeseer has a higher δ_{avg}-hyperbolicity, HAT performs not well in Citeseer with 64-dimension. Nonetheless, HAT also has competitive results. Moreover, we can find that the GNN-based methods usually perform better than other baselines (i.e., DeepWalk and PoincaréEmb), because of combining the graph structure and node features in their models. Furthermore, compared to Euclidean GNNs, HAT performs better in most cases, especially in low dimension, suggesting the superiority of modeling graph in hyperbolic spaces.

More detailed introduction of HAT can be found in [242].

6.3   Lorentzian Graph Convolutional Network

6.3.1   Overview

Most GNNs learn the node representations in Euclidean geometry, but that could have a high distortion in the case of embedding graphs with scale-free or hierarchical structure. Recently, some GNNs are proposed to deal with this problem in non-Euclidean geometry, e.g., hyperbolic geometry.

Although hyperbolic GNNs achieve promising performance, existing hyperbolic graph operations actually cannot rigorously follow the hyperbolic geometry, which may limit the ability of hyperbolic geometry and thus hurt the performance of hyperbolic GNNs. In this paper, we propose a novel hyperbolic GNN, i.e., Lorentzian graph convolutional network (LGCN), which rigorously guarantees the learned node features follow the hyperbolic geometry. Specifically, we rebuild the graph operations of hyperbolic GCNs with Lorentzian version, e.g., the feature transformation and non-linear activation. Also, an elegant neighborhood aggregation method is designed based on the centroid of Lorentzian distance. Experiments on six datasets show that LGCN performs better than the state-of-the-art methods. LGCN has lower distortion to learn the representation of tree-likeness graphs compared with existing hyperbolic GCNs.

6.3.2   The LGCN Method

6.3.2.1 The LGCN Framework

In this section, we propose LGCN which designs graph operations to guarantee the mathematical meanings in hyperbolic spaces. Specifically, LGCN first maps the input node features into hyperbolic spaces and then conducts feature transformation via a delicately designed Lorentzian matrix-vector multiplication. Also, the centroid-based Lorentzian aggregation is proposed to aggregate features, and the aggregation weights are learned by a self-attention mechanism. Moreover, Lorentzian pointwise non-linear activation is followed to obtain the output node features. Please note that LGCN leverages the hyperboloid model to describe hyperbolic spaces, which is defined as the n-dimensional hyperboloid manifold with constant negative curvature $-1/\beta$ ($\beta > 0$): $\mathbb{H}^{n,\beta} := \{\mathbf{x} \in \mathbb{R}^{n+1} : \langle \mathbf{x}, \mathbf{x} \rangle_{\mathcal{L}} = -\beta, x_0 > 0\}$, where $\langle \mathbf{x}, \mathbf{y} \rangle_{\mathcal{L}} := -x_0 y_0 + \sum_{i=1}^{n} x_i y_i$. is the Lorentzian scalar product. The details of LGCN are introduced in the following.

6.3.2.2 Mapping Feature with Different Curvature

The input node features of LGCN could live in the Euclidean spaces or hyperbolic spaces. For k-dimensional input features, we denote them as $\mathbf{h}^{k,E} \in \mathbb{R}^k$ (E indicates Euclidean spaces) and $\mathbf{h}^{k,\beta'} \in \mathbb{H}^{k,\beta'}$, respectively. If original features live in Euclidean spaces, we need to map them into hyperbolic spaces. We assume that the input features $\mathbf{h}^{k,E}$ live in the tangent space of $\mathbb{H}^{k,\beta}$ at its origin $\mathbf{0} = (\sqrt{\beta}, 0, \dots, 0) \in \mathbb{H}^{k,\beta}$, i.e., $\mathcal{T}_0 \mathbb{H}^{k,\beta}$. A "0" element is added at the first coordinate of $\mathbf{h}^{k,E}$ to satisfy the constraint $\langle (0, \mathbf{h}^{k,E}), \mathbf{0} \rangle_{\mathcal{L}} = 0$ in tangent spaces, i.e., $\mathcal{T}_\mathbf{x} \mathbb{H}^{n,\beta} := \{\mathbf{v} \in \mathbb{R}^{n+1} : \langle \mathbf{v}, \mathbf{x} \rangle_{\mathcal{L}} = 0\}$. Thus, the input feature $\mathbf{h}^{k,E} \in \mathbb{R}^k$ can be mapped to the hyperbolic spaces via exponential map:

$$\mathbf{h}^{k,\beta} = \exp_{\mathbf{0}}^{\beta} \left((0, \mathbf{h}^{k,E}) \right), \exp_{\mathbf{x}}^{\beta}(\mathbf{v}) = \cosh\left(\frac{\|\mathbf{v}\|_{\mathcal{L}}}{\sqrt{\beta}} \right) \mathbf{x} + \sqrt{\beta} \sinh\left(\frac{\|\mathbf{v}\|_{\mathcal{L}}}{\sqrt{\beta}} \right) \frac{\mathbf{v}}{\|\mathbf{v}\|_{\mathcal{L}}}. \quad (6.8)$$

If the input features $\mathbf{h}^{k,\beta'}$ live in a hyperbolic space (e.g., the output of previous LGCN layer), whose curvature $-1/\beta'$ might be different from the curvature of current hyperboloid model. We can transform it into the hyperboloid model with a specific curvature $-1/\beta$:

$$\mathbf{h}^{k,\beta} = \exp_{\mathbf{0}}^{\beta}(\log_{\mathbf{0}}^{\beta'}(\mathbf{h}^{k,\beta'})), \log_{\mathbf{x}}^{\beta}(\mathbf{y}) = d_{\mathbb{H}}^{\beta}(\mathbf{x},\mathbf{y}) \frac{\mathbf{y} + \frac{1}{\beta}\langle \mathbf{x},\mathbf{y}\rangle_{\mathcal{L}}\mathbf{x}}{\|\mathbf{y} + \frac{1}{\beta}\langle \mathbf{x},\mathbf{y}\rangle_{\mathcal{L}}\mathbf{x}\|_{\mathcal{L}}}. \tag{6.9}$$

6.3.2.3 Lorentzian Feature Transformation

In order to apply *linear transformation* on the hyperboloid model, we propose the Lorentzian matrix-vector multiplication:

Definition (*Lorentzian matrix-vector multiplication*) If $\mathbf{M} : \mathbb{R}^n \to \mathbb{R}^m$ is a linear map with matrix representation, given two points $\mathbf{x} = (x_0, \ldots, x_n) \in \mathbb{H}^{n,\beta}$, $\mathbf{v} = (v_0, \ldots, v_n) \in \mathcal{T}_{\mathbf{0}}\mathbb{H}^{n,\beta}$, we have $\mathbf{M}^{\otimes^{\beta}}(\mathbf{x}) = \exp_{\mathbf{0}}^{\beta}(\hat{\mathbf{M}}(\log_{\mathbf{0}}^{\beta}(\mathbf{x})))$, $\hat{\mathbf{M}}(\mathbf{v}) = (0, \mathbf{M}(v_1, \ldots, v_n))$. $\qquad\square$

A key difference between Lorentzian matrix-vector multiplication and other matrix-vector multiplications on the hyperboloid model [21, 120] is the size of the matrix $\mathbf{M}$. Assuming a n-dimensional feature needs to be transformed into a m-dimensional feature. Naturally, the size of matrix $\mathbf{M}$ should be $m \times n$, which is satisfied by Lorentzian matrix-vector multiplication. However, the size of matrix $\mathbf{M}$ is $(m+1) \times (n+1)$ for other methods [21, 120]. Moreover, the Lorentzian matrix-vector multiplication has the following property:

Theorem *Given a point in hyperbolic space, which is represented by $\mathbf{x}^{n,\beta} \in \mathbb{H}^{n,\beta}$ using hyperboloid model or $\mathbf{x}^{n,\alpha} \in \mathbb{D}^{n,\alpha}$ using Poincaré ball model [59], respectively. Let $\mathbf{M}$ be a $m \times n$ matrix, Lorentzian matrix-vector multiplication $\mathbf{M}\otimes^{\beta}\mathbf{x}^{n,\beta}$ used in hyperboloid model is equivalent to Möbius matrix-vector multiplication $\mathbf{M}\otimes^{\alpha}\mathbf{x}^{n,\alpha}$ used in Poincaré ball model.* $\qquad\square$

This property elegantly bridges the relation between the hyperboloid model and Poincaré ball model w.r.t. matrix-vector multiplication. We use the Lorentzian matrix-vector multiplication to conduct feature transformation on the hyperboloid model as $\mathbf{h}^{d,\beta} = \mathbf{M}\otimes^{\beta}\mathbf{h}^{k,\beta}$.

6.3.2.4 Lorentzian Neighborhood Aggregation

In Euclidean spaces, the neighborhood aggregation is to compute the weight arithmetic mean or centroid (also called center of mass) of its neighborhood features (see Fig. 6.3a). Therefore, we aim to aggregate neighborhood features in hyperbolic spaces to follow these meanings. Fréchet mean [53, 97] provides a feasible way to compute the centroid in Riemannian manifold. Also, the arithmetic mean can be interpreted as a kind of Fréchet mean. Thus, Fréchet mean meets the meanings of neighborhood aggregation. The main idea of Fréchet mean is to minimize an expectation of (squared) distances with a set of points. However, Fréchet mean does not have a closed form solution w.r.t. the intrinsic distance $d_{\mathbb{H}}^{\beta}$ in hyperbolic spaces, and it has to be inefficiently computed by gradient descent. Therefore, we propose an elegant neighborhood aggregation method based on the centroid of the squared Lorentzian distance, which can well balance the mathematical meanings and efficiency:

Theorem (Lorentzian aggregation via centroid) *For a node feature* $\mathbf{h}_i^{d,\beta} \in \mathbb{H}^{d,\beta}$, *a set of its neighborhoods* $N(i)$ *with aggregation weights* $w_{ij} > 0$, *the neighborhood aggregation consists in the centroid* $\mathbf{c}^{d,\beta}$ *of nodes, which minimizes the problem:* $\arg\min_{\mathbf{c}^{d,\beta} \in \mathbb{H}^{d,\beta}} \sum_{j \in N(i) \cup \{i\}} w_{ij} d_{\mathcal{L}}^2(\mathbf{h}_j^{d,\beta}, \mathbf{c}^{d,\beta})$, *where* $d_{\mathcal{L}}^2(\cdot, \cdot)$ *denotes squared Lorentzian distance, and this problem has closed form solution:*

$$\mathbf{c}^{d,\beta} = \sqrt{\beta} \frac{\sum_{j \in N(i) \cup \{i\}} w_{ij} \mathbf{h}_j^{d,\beta}}{\left| \| \sum_{j \in N(i) \cup \{i\}} w_{ij} \mathbf{h}_j^{d,\beta} \|_{\mathcal{L}} \right|}. \tag{6.10}$$

For points $\mathbf{x}^{n,\beta}, \mathbf{y}^{n,\beta} \in \mathbb{H}^{n,\beta}$, the squared Lorentzian distance is defined as [157]: $d_{\mathcal{L}}^2(\mathbf{x}^{n,\beta}, \mathbf{y}^{n,\beta}) = -2\beta - 2\langle \mathbf{x}^{n,\beta}, \mathbf{y}^{n,\beta} \rangle_{\mathcal{L}}$.

Figure 6.3b illustrates Lorentzian aggregation via centroid. Similar to Fréchet/Karcher means, the node features computed by Lorentzian aggregation are the minimum of an expectation of squared Lorentzian distance. Also, the features of aggregation in Lorentzian neigh-

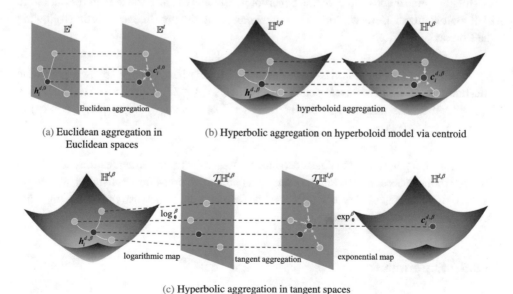

(a) Euclidean aggregation in   (b) Hyperbolic aggregation on hyperboloid model via centroid
Euclidean spaces

(c) Hyperbolic aggregation in tangent spaces

Fig. 6.3 Three types of neighborhood aggregation. The three types of aggregation can be considered as computing centroids in Euclidean spaces, hyperbolic spaces, tangent spaces, respectively

borhood aggregation are the centroids in the hyperboloid model in geometry [109, 157]. On the other hand, some hyperbolic GCNs [21, 120, 242] aggregate neighborhoods in tangent spaces (as shown in Fig. 6.3c), that can only be regarded as centroid or arithmetic mean in the tangent spaces, rather than hyperbolic spaces. Thus Lorentzian aggregation via centroid of squared Lorentzian distance is a promising method, which satisfies more elegant mathematical meanings compared to other hyperbolic GCNs.

As shown in Eq. (6.10), there is an aggregation weight w_{ij} indicating the importance of neighborhoods for a center node. Here we propose a self-attention mechanism to learn the aggregation weights w_{ij}. For two-node features $\mathbf{h}_i^{d,\beta}, \mathbf{h}_i^{d,\beta} \in \mathbb{H}^{d,\beta}$, the attention coefficient μ_{ij}, which indicates the importance of node j to node i, can be computed as $\mu_{ij} = ATT(\mathbf{h}_i^{d,\beta}, \mathbf{h}_j^{d,\beta}, \mathbf{M}_{att})$, where $ATT(\cdot)$ indicates the function of computing the attention coefficient and the $d \times d$ matrix $\mathbf{M}_{att}$ is to transform the node features into attention-based ones. Considering a large attention coefficient μ_{ij} represents a high similarity of nodes j and i, we define $ATT(\cdot)$ based on squared Lorentzian distance, as $\mu_{ij} = -d_{\mathcal{L}}^2(\mathbf{M}_{att} \otimes^\beta \mathbf{h}_i^{d,\beta}, \mathbf{M}_{att} \otimes^\beta \mathbf{h}_j^{d,\beta})$. For all the neighbors $N(i)$ of node i (including itself), we normalize them using the softmax function to compute the aggregation weight:
$$w_{ij} = \frac{\exp(\mu_{ij})}{\sum_{t \in N(i) \cup \{i\}} \exp(\mu_{it})}.$$

6.3.2.5 Lorentzian Pointwise Non-linear Activation

Non-linear activation is an indispensable part of GCNs. Similar to feature transformation, existing non-linear activations on the hyperboloid model [21] also make features out of the hyperboloid model. Here, we derive the Lorentzian pointwise non-linear activation following the Lorentzian version:

Definition (*Lorentzian pointwise non-linear activation*) If $\sigma : \mathbb{R}^n \to \mathbb{R}^n$ is a pointwise non-linearity map, given two points $\mathbf{x} = (x_0, \ldots, x_n) \in \mathbb{H}^{n,\beta}$ and $\mathbf{v} = (v_0, \ldots, v_n) \in \mathcal{T}_0\mathbb{H}^{n,\beta}$, the Lorentzian version $\sigma^{\otimes^\beta}$ is $\sigma^{\otimes^\beta}(\mathbf{x}) = \exp_0^\beta(\hat{\sigma}^{\otimes^\beta}(\log_0^\beta(\mathbf{x})))$, $\hat{\sigma}^{\otimes^\beta}(\mathbf{v}) = (0, \sigma(v_1), \ldots, \sigma(v_n)))$. $\qquad\square$

The Lorentzian pointwise non-linear activation ensures the transformed features still live in the hyperbolic spaces. Following the Lorentzian pointwise non-linear activation, the output of the LGCN layer is $\mathbf{u}^{d,\beta} = \sigma^{\otimes^\beta}(\mathbf{c}^{d,\beta})$, which can be used to downstream tasks, e.g., link prediction and node classification.

6.3.3   Experiments

6.3.3.1 Experimental Settings

Datasets. We utilize six datasets in our experiments: Cora, Citeseer, Pubmed [221], Amazon [128, 165], USA [159], and Disease [21]. We compute δ_{avg}-hyperbolicity [3] to quantify the tree-likeliness of these datasets. A low δ_{avg}-hyperbolicity of a graph indicates that it has an underlying hyperbolic geometry.

Baselines. We compare our method with the following state-of-the-art methods: (1) A Euclidean network embedding model, i.e., DeepWalk [154] and a hyperbolic network embedding model, i.e., PoincaréEmb [141]; (2) Euclidean GCNs, i.e., GCN [102], GAT [189]; (3) Hyperbolic GCNs, i.e., HGCN [21], HAT [242].

6.3.3.2 Link Prediction

We compute the probability scores for edges by leveraging the Fermi-Dirac decoder [21, 105, 141]. For the output node features $\mathbf{u}_i^{d,\beta}$ and $\mathbf{u}_j^{d,\beta}$, the probability of existing the edge e_{ij} between $\mathbf{u}_i^{d,\beta}$ and $\mathbf{u}_j^{d,\beta}$ is given as $p(\mathbf{u}_i^{d,\beta}, \mathbf{u}_i^{d,\beta}) = 1/(e^{(d_{\mathcal{L}}^2(\mathbf{u}_i^{d,\beta}, \mathbf{u}_i^{d,\beta})-r)/t} + 1)$, where r and t are hyper-parameters. We then minimize the cross-entropy loss to train the LGCN model. Following [21], the edges are split into 85, 5, 10% randomly for training, validation and test sets for all datasets, and the evaluation metric is AUC.

The results are shown in Table 6.3. We can see that LGCN performs best in all cases, and its superiority is more significant for the low dimension setting. Suggesting the graph operations of LGCN provide powerful ability to embed graphs. Moreover, hyperbolic GCNs perform better than Euclidean GCNs for datasets with lower δ_{avg}, which further confirms the capability of hyperbolic spaces in modeling tree-likeness graph data. Furthermore, compared with network embedding methods, GCNs achieve better performance in most cases, which indicates GCNs can benefit from both structure and feature information in a graph.

6.3.3.3 Node Classification

Here we evaluate the performance of LGCN on the node classification task. We split nodes in the disease dataset into 30/10/60% for training, validation and test sets [21]. For the other datasets, we use only 20 nodes per class for training, 500 nodes for validation, 1000 nodes for test. The settings are same with [21, 102, 189, 221]. The accuracy is used to evaluate the results.

Table 6.4 reports the performance. We can observe similar results to Table 6.3. That is, LGCN performs better than the baselines in most cases. Also, hyperbolic GCNs outperform Euclidean GCNs for datasets with lower δ_{avg}, and GCNs perform better than network embedding methods. Moreover, we notice that hyperbolic GCNs do not have an obvious advantage compared with Euclidean GCNs on Citeseer dataset, which has the biggest δ_{avg}. We think no obvious tree-likeness structure of Citeseer makes those hyperbolic GCNs not work well on this task. In spite of this, benefiting from the well-defined Lorentzian graph operations, LGCN also achieves very competitive results.

The complete method with more experiments can be found in [243].

Table 6.3 AUC (%) for link prediction task. The best results are marked by bold numbers

Dataset	dimension	DeepWalk	PoincaréEmb	GCN	GAT	HGCN	HAT	LGCN
Disease $\delta_{avg} = 0.00$	8	57.3±1.0	67.9±1.1	76.9±0.8	73.5±0.8	84.1±0.7	83.9±0.7	**89.2±0.7**
	16	55.2±1.7	70.9±1.0	78.2±0.7	73.8±0.6	91.2±0.6	91.8±0.5	**96.6±0.6**
	32	49.1±1.3	75.1±0.7	78.7±0.5	75.7±0.3	91.8±0.3	92.3±0.5	**96.3±0.5**
	64	47.3±0.1	76.3±0.3	79.8±0.5	77.9±0.3	92.7±0.4	93.4±0.4	**96.8±0.4**
USA $\delta_{avg} = 0.16$	8	91.5±0.1	92.3±0.2	89.0±0.6	89.6±0.9	91.6±0.8	92.7±0.8	**95.3±0.2**
	16	92.3±0.0	93.6±0.2	90.2±0.5	91.1±0.5	93.4±0.3	93.6±0.6	**96.3±0.2**
	32	92.5±0.1	94.5±0.1	90.7±0.5	91.7±0.5	93.9±0.2	94.2±0.6	**96.5±0.1**
	64	92.5±0.1	95.5±0.1	91.2±0.3	93.3±0.4	94.2±0.2	94.6±0.6	**96.4±0.2**
Amazon $\delta_{avg} = 0.20$	8	96.1±0.0	95.1±0.4	91.1±0.6	91.3±0.6	93.5±0.6	94.8±0.8	**96.4±1.1**
	16	96.6±0.0	96.7±0.3	92.8±0.8	92.8±0.9	96.3±0.9	96.9±1.0	**97.3±0.8**
	32	96.4±0.0	96.7±0.1	93.3±0.9	95.1±0.5	97.2±0.8	97.1±0.7	**97.5±0.3**
	64	95.9±0.0	97.2±0.1	94.6±0.8	96.2±0.2	97.1±0.7	97.3±0.6	**97.6±0.5**
Cora $\delta_{avg} = 0.35$	8	86.9±0.1	84.5±0.7	87.8±0.9	87.4±1.0	91.4±0.5	91.1±0.4	**92.0±0.5**
	16	85.3±0.8	85.8±0.8	90.6±0.7	93.2±0.4	93.1±0.4	93.0±0.3	**93.6±0.4**
	32	82.3±0.4	86.5±0.6	92.0±0.6	93.6±0.3	93.3±0.3	93.1±0.3	**94.0±0.4**
	64	81.6±0.4	86.7±0.5	92.8±0.4	93.5±0.3	93.5±0.2	93.3±0.3	**94.4±0.2**
Pubmed $\delta_{avg} = 0.36$	8	81.1±0.1	83.3±0.5	86.8±0.7	87.0±0.8	94.6±0.2	94.4±0.3	**95.4±0.2**
	16	81.2±0.1	85.1±0.5	90.9±0.6	91.6±0.3	96.1±0.2	96.2±0.3	**96.6±0.1**
	32	76.4±0.1	86.5±0.1	93.2±0.5	93.6±0.2	96.2±0.2	96.3±0.2	**96.8±0.1**
	64	75.3±0.1	87.4±0.1	93.6±0.4	94.6±0.2	96.5±0.2	96.5±0.1	**96.9±0.0**
Citeseer $\delta_{avg} = 0.46$	8	80.7±0.3	79.2±1.0	90.3±1.2	89.5±0.9	93.2±0.5	93.1±0.3	**93.9±0.6**
	16	78.5±0.5	79.7±0.7	92.9±0.7	92.2±0.7	94.3±0.4	93.6±0.5	**95.4±0.5**
	32	73.1±0.4	79.8±0.6	94.3±0.6	93.4±0.4	94.7±0.3	94.2±0.5	**95.8±0.3**
	64	72.3±0.3	79.6±0.6	95.4±0.5	94.4±0.3	94.8±0.3	94.3±0.2	**96.4±0.2**

6.4   Hyperbolic Heterogeneous Graph Representation

6.4.1   Overview

Heterogeneous Graph (HG) embedding has attracted considerable research attention recently. Most HG embedding methods choose Euclidean spaces to represent HG graphs, which is because Euclidean spaces are the natural generalization of our intuition-friendly, and visual three-dimensional space. However, a fundamental problem is that what are the appropriate or intrinsic underlying spaces of HGs? Therefore, we wonder whether Euclidean spaces are the intrinsic spaces of HGs?

Recently, hyperbolic spaces have gained momentum in the context of network science [105]. Hyperbolic spaces are spaces of constant negative curvature [17]. A superiority of hyperbolic spaces is that they expand faster than Euclidean spaces [141]. Therefore, it is easy to model complex data with low-dimensional embedding in hyperbolic spaces. Due to the characteristic of hyperbolic spaces, [105] assumes hyperbolic spaces underlie complex network and finds that data with power-law structure is suitable to be modeled in hyper-

Table 6.4 Accuracy (%) for node classification task. The best results are marked by bold numbers

Dataset	dimension	sDeepWalk	PoincaréEmb	GCN	GAT	HGCN	HAT	LGCN
Disease $\delta_{avg} = 0.00$	8	59.6±1.6	57.0±0.8	75.1±1.1	76.7±0.7	81.5±1.3	82.3±1.2	**82.9±1.2**
	16	61.5±2.2	56.1±0.7	78.3±1.0	76.6±0.8	82.8±0.8	83.6±0.9	**84.4±0.8**
	32	62.0±0.3	58.7±0.7	81.0±0.9	79.3±0.7	84.0±0.8	84.9±0.9	**86.8±0.8**
	64	61.8±0.5	60.1±0.8	82.7±0.9	80.4±0.7	84.3±0.8	85.1±0.8	**87.1±0.8**
USA $\delta_{avg} = 0.16$	8	44.3±0.6	38.9±1.1	50.5±0.5	47.8±0.7	50.5±1.1	50.7±1.0	**51.6±1.1**
	16	42.3±1.3	38.3±1.0	50.9±0.6	49.5±0.7	51.1±1.0	51.3±0.9	**51.9±0.9**
	32	39.0±1.0	39.0±0.8	50.6±0.5	49.1±0.6	51.2±0.9	51.5±0.8	**52.4±0.9**
	64	42.7±0.8	39.2±0.8	51.1±0.6	49.6±0.6	52.4±0.8	52.5±0.7	**52.8±0.8**
Amazon $\delta_{avg} = 0.20$	8	66.7±1.0	65.3±1.1	70.9±1.1	70.0±0.9	71.7±1.3	71.0±1.0	**72.0±1.3**
	16	67.5±0.8	67.0±0.7	70.9±1.1	72.7±0.8	72.7±1.3	73.3±1.0	**75.0±1.1**
	32	70.0±0.5	68.1±0.3	71.5±0.8	72.5±0.7	75.3±1.0	74.9±0.8	**75.5±0.9**
	64	70.3±0.7	67.3±0.4	73.0±0.6	72.9±0.8	75.5±0.6	75.4±0.7	**75.8±0.6**
Cora $\delta_{avg} = 0.35$	8	64.5±1.2	57.5±0.6	80.3±0.8	80.4±0.8	80.0±0.7	**82.8±0.7**	82.6±0.8
	16	65.2±1.6	64.4±0.3	81.9±0.6	81.7±0.7	81.3±0.6	83.1±0.6	**83.3±0.7**
	32	65.9±1.5	64.9±0.4	81.5±0.4	82.6±0.7	81.7±0.7	83.2±0.6	**83.5±0.6**
	64	66.5±1.7	68.6±0.4	81.6±0.4	83.1±0.6	82.1±0.7	83.1±0.5	**83.5±0.5**
Pubmed $\delta_{avg} = 0.36$	8	73.2±0.7	66.0±0.8	78.6±0.4	71.9±0.7	77.9±0.6	78.5±0.6	**78.8±0.5**
	16	73.9±0.8	68.0±0.4	79.1±0.5	75.9±0.7	78.4±0.4	78.6±0.5	**78.6±0.7**
	32	72.4±1.0	68.4±0.5	78.7±0.5	78.2±0.6	78.6±0.6	78.8±0.6	**78.9±0.6**
	64	73.5±1.0	69.9±0.6	79.1±0.5	78.7±0.4	79.3±0.5	79.0±0.6	**79.6±0.6**
Citeseer $\delta_{avg} = 0.46$	8	47.8±1.6	38.6±0.4	68.9±0.7	69.5±0.8	70.9±0.6	71.2±0.7	**71.8±0.7**
	16	46.2±1.5	40.4±0.5	70.2±0.6	71.6±0.7	71.2±0.5	**71.9±0.6**	71.9±0.7
	32	43.6±1.9	43.5±0.5	70.4±0.5	**72.6±0.7**	71.9±0.4	72.4±0.5	72.5±0.5
	64	46.6±1.4	43.6±0.4	70.8±0.4	72.4±0.7	71.7±0.5	72.2±0.5	**72.5±0.6**

bolic spaces. Also, some researchers begin to embed different data in hyperbolic spaces. For instance, [58, 141] learn the hyperbolic embeddings of homogeneous networks. However, it is unknown whether HGs are suitable to be embedded in hyperbolic spaces.

In this section, we analyze the relation distribution in HGs and propose HHNE, which is able to preserve the structure and semantic information in hyperbolic spaces. HHNE leverages the meta-path guided random walk to generate heterogeneous neighborhoods to capture the structure and semantic relations in HGs. Then the proximity between nodes is measured by the distance in hyperbolic spaces. Also, HHNE is able to maximize the proximity between the neighborhood nodes while minimize the proximity between the negative sampled nodes. The optimization strategy of HHNE is derived to optimize hyperbolic embeddings iteratively.

6.4.2   The HHNE Method

6.4.2.1 The HHNE Framework

HHNE leverages the meta-path guided random walk to obtain neighbors for each node to capture the structure and semantic relations in HGs. Also, HHNE learns the embeddings by maximizing the proximity between the neighborhood nodes and minimizing the proximity between the negative sampled nodes. Moreover, we derive the optimization strategy of HHNE to upgrade the hyperbolic embeddings.

6.4.2.2 Hyperbolic HG Embedding

To design HG embedding methods in hyperbolic spaces, we makes use of the Poincaré ball model to describe hyperbolic spaces. Let $\mathbb{D}^d = \{x \in \mathbb{R}^d : \|x\| < 1\}$ be the *open d-dimensional unit ball*. The Poincaré ball model is defined by the manifold $\mathbb{D}^d$ equipped with the following Riemannian metric tensor $g_x^{\mathbb{D}}$:

$$g_x^{\mathbb{D}} = \lambda_x^2 g^{\mathbb{E}} \quad \text{where } \lambda_x := \frac{2}{1 - \|x\|^2}, \tag{6.11}$$

where $x \in \mathbb{D}^d$, $g^{\mathbb{E}} = \mathbf{I}$ denotes the Euclidean metric tensor.

HHNE aims to learn the representation of nodes to preserve the structure and semantic correlations in hyperbolic spaces. Given an HG $G = (V, E, T, \phi, \psi)$ with $|T_V| > 1$, HHNE is interested in learning the embeddings $\Theta = \{\theta_i\}_{i=1}^{|V|}$, $\theta_i \in \mathbb{D}^d$. HHNE preserves the structure by facilitating the proximity between a node and its neighborhoods. HHNE uses meta-path guided random walks [42] to obtain heterogeneous neighborhoods of a node. In meta-path guided random walks, the node sequences are restrained by the node types which are defined by meta-paths. Specifically, let t_{v_i} and t_{e_i} as the types of node v_i and edge e_i, respectively, given a meta-path $\mathcal{P} = t_{v_1} \xrightarrow{t_{e_1}} \dots t_{v_i} \xrightarrow{t_{e_i}} \dots \xrightarrow{t_{e_{n-1}}} t_{v_n}$, the transition probability at step i is defined as follows:

$$p(v^{i+1}|v_{t_{v_i}}^i, \mathcal{P}) = \begin{cases} \frac{1}{|N_{t_{v_{i+1}}}(v_{t_{v_i}}^i)|} & (v^{i+1}, v_{t_{v_i}}^i) \in E, \phi(v^{i+1}) = t_{v_{i+1}} \\ 0 & \text{otherwise,} \end{cases} \tag{6.12}$$

where $v_{t_{v_i}}^i$ is node $v \in V$ with type t_{v_i}, and $N_{t_{v_{i+1}}}(v_{t_{v_i}}^i)$ denotes the $t_{v_{i+1}}$ type of neighborhood of node $v_{t_{v_i}}^i$. The meta-path guided random walk strategy ensures that the semantic relationships between different types of nodes can be properly incorporated into HHNE.

In order to preserve the proximity between nodes and its neighborhoods in hyperbolic spaces, HHNE uses distances in Poincaré ball model to measure their proximity. Given nodes embeddings $\theta_i, \theta_j \in \mathbb{D}^d$, the distance in Poincaré ball is given by:

$$d_{\mathbb{D}}(\theta_i, \theta_j) = \cosh^{-1}\left(1 + 2\frac{\|\theta_i - \theta_j\|^2}{(1 - \|\theta_i\|^2)(1 - \|\theta_j\|^2)}\right). \tag{6.13}$$

It is worth noting that as the Poincaré ball model is defined in metric spaces, the distance in Poincaré ball meets the triangle inequality and can well preserve the transitivity in HG. Then, HHNE uses a probability to measure the node c_t is a neighborhood of node v as follows:

$$p(v|c_t; \Theta) = \sigma[-d_{\mathbb{D}}(\theta_v, \theta_{c_t})],$$

where $\sigma(x) = \frac{1}{1+\exp(-x)}$. Then the object of HHNE is to maximize the probability as follows:

$$\arg\max_{\Theta} \sum_{v \in V} \sum_{c_t \in C_t(v)} \log p(v|c_t; \Theta). \tag{6.14}$$

To achieve efficient optimization, HHNE leverages the negative sampling proposed in [132], which basically samples a small number of negative objects to enhance the influence of positive objects. For a given node v, HHNE aims to maximize the proximity between v and its neighborhood c_t while minimizes the proximity between v and its negative sampled node n. Therefore, the objective function Eq. (6.14) can be rewritten as follows:

$$\mathcal{L}(\Theta) = \log \sigma[-d_{\mathbb{D}}(\theta_{c_t}, \theta_v)] + \sum_{m=1}^{M} \mathbb{E}_{n^m \sim P(n)}\{\log \sigma[d_{\mathbb{D}}(\theta_{n^m}, \theta_v)]\}, \tag{6.15}$$

where $P(n)$ is a pre-defined distribution from which a negative node n^m is drew from for M times. HHNE builds the node frequency distribution by drawing nodes regardless of their types.

6.4.2.3 Optimization

As the parameters of the model live in a Poincaré ball which has a Riemannian manifold structure, the back-propagated gradient is a Riemannian gradient. It means that the Euclidean gradient-based optimization, such as $\theta_i \leftarrow \theta_i + \eta \nabla_{\theta_i}^E \mathcal{L}(\Theta)$, makes no sense as an operation in the Poincaré ball, because the addition operation is not defined in this manifold. Instead, HHNE can optimize Eq. (6.15) via a Riemannian stochastic gradient descent (RSGD) optimization method [10]. In particular, let $\mathcal{T}_{\theta_i} \mathbb{D}^d$ denote the tangent space of a node embedding $\theta_i \in \mathbb{D}^d$, and HHNE can compute the Riemannian gradient $\nabla_{\theta_i}^R \mathcal{L}(\Theta) \in \mathcal{T}_{\theta_i} \mathbb{D}^d$ of $\mathcal{L}(\Theta)$. Using RSGD, HHNE can be optimized by maximizing Eq. (6.15), and a node embedding can be updated in the form of:

$$\theta_i \leftarrow \exp_{\theta_i}(\eta \nabla_{\theta_i}^R \mathcal{L}(\Theta)), \tag{6.16}$$

where $\exp_{\theta_i}(\cdot)$ is exponential map in the Poincaré ball. The exponential map is given by [58]:

$$\exp_{\theta_i}(s) = \frac{\lambda_{\theta_i}\left(\cosh(\lambda_{\theta_i}\|s\|) + \langle\theta_i, \frac{s}{\|s\|}\rangle \sinh(\lambda_{\theta_i}\|s\|)\right)}{1 + (\lambda_{\theta_i} - 1)\cosh(\lambda_{\theta_i}\|s\|) + \lambda_{\theta_i}\langle\theta_i, \frac{s}{\|s\|}\rangle\sinh(\lambda_{\theta_i}\|s\|)}\theta_i$$

$$+ \frac{\frac{1}{\|s\|}\sinh(\lambda_{\theta_i}\|s\|)}{1 + (\lambda_{\theta_i} - 1)\cosh(\lambda_{\theta_i}\|s\|) + \lambda_{\theta_i}\langle\theta_i, \frac{s}{\|s\|}\rangle\sinh(\lambda_{\theta_i}\|s\|)}s.$$

(6.17)

As the Poincaré ball model is a conformal model of hyperbolic spaces, i.e., $g_x^{\mathbb{D}} = \lambda_x^2 g^{\mathbb{E}}$, the Riemannian gradient ∇^R is obtained by rescaling the Euclidean gradient ∇^E by the inverse of the metric tensor, i.e., $\frac{1}{g_x^{\mathbb{D}}}$:

$$\nabla_{\theta_i}^R \mathcal{L} = \left(\frac{1}{\lambda_{\theta_i}}\right)^2 \nabla_{\theta_i}^E \mathcal{L}.$$

(6.18)

Furthermore, the gradients of Eq. (6.15) can be derived as follows:

$$\frac{\partial \mathcal{L}}{\partial \theta_{u^m}} = \frac{4}{\alpha\sqrt{\gamma^2 - 1}}\left[\mathbb{I}_v[u^m] - \sigma(-d_{\mathbb{D}}(\theta_{c_t}, \theta_{u^m}))\right]$$
$$\cdot \left[\frac{\theta_{c_t}}{\beta_m} - \frac{\|\theta_{c_t}\|^2 - 2\langle\theta_{c_t}, \theta_{u^m}\rangle + 1}{\beta_m^2}\theta_{u^m}\right],$$

(6.19)

$$\frac{\partial \mathcal{L}}{\partial \theta_{c_t}} = \sum_{m=0}^{M}\frac{4}{\beta_m\sqrt{\gamma^2 - 1}}\left[\mathbb{I}_v[u^m] - \sigma(-d_{\mathbb{D}}(\theta_{c_t}, \theta_{u^m}))\right]$$
$$\cdot \left[\frac{\theta_{u^m}}{\alpha} - \frac{\|\theta_{u^m}\|^2 - 2\langle\theta_{c_t}, \theta_{u^m}\rangle + 1}{\alpha^2}\theta_{c_t}\right],$$

(6.20)

where $\alpha = 1 - \|\theta_{c_t}\|^2$, $\beta_m = 1 - \|\theta_{u^m}\|^2$, $\gamma = 1 + \frac{2}{\alpha\beta}\|\theta_{c_t} - \theta_{u^m}\|^2$ and when $m = 0, u^0 = v$. $\mathbb{I}_v[u]$ is an indicator function to indicate whether u is v. Then, HHNE can be updated by using Eqs. (6.19)–(6.20) iteratively.

6.4.3   Experiments

6.4.3.1 Experimental Settings
Datasets. The basic statistics of the two HGs used in our experiments are shown in Table 6.5.

Table 6.5 Statistics of datasets

DBLP	# Author (A)	# Paper (P)	# Venue (V)	# P-A	# P-V
	14475	14376	20	41794	14376
MovieLens	# Actor (A)	# Movie (M)	# Director (D)	# M-A	# M-D
	11718	9160	3510	64051	9160

Baselines. HHNE is compared with the following state-of-the-art methods: (1) the homogeneous graph embedding methods, i.e., DeepWalk [154], LINE [181]; (2) the heterogeneous graph embedding methods, i.e., metapath2vec [42]; (3) the hyperbolic homogeneous graph embedding methods, i.e., PoincaréEmb [141].

6.4.3.2 Network Reconstruction

A good HG embedding method should ensure that the learned embeddings can preserve the original HG structure. The reconstruction error in relation to the embedding dimension is then a measure for the capacity of the model. More specifically, we use network embedding methods to learn feature representations. Then for each type of links in the HG, we enumerate all pairs of objects that can be connected by such a link and calculate their proximity [87], i.e., the distance in Poincaré ball model for HHNE and PoincaréEmb. Finally, we use the AUC [49] to evaluate the performance of each embedding method. For example, for link type "write", we calculate all pairs of authors and papers in DBLP and compute the proximity for each pair. Then using the links between authors and papers in real DBLP network as ground truth, we compute the AUC value for each embedding method.

The results are shown in Table 6.6. As we can see, HHNE consistently performs the best in all the tested HGs. The results demonstrate that HHNE can effectively preserve the original network structure and reconstruct the network, especially on the reconstruction of P-V and M-D edges. Also, please note that HHNE achieves very promising results when the embedding dimension is very small. This suggests that regarding hyperbolic spaces underlying HG is reasonable and hyperbolic spaces have strong ability of modeling network when the dimension of spaces is small.

6.4.3.3 Link Prediction

Link prediction aims to infer the unknown links in an HG given the observed HG structure, which can be used to test the generalization performance of a network embedding method. For each type of edge, 20% of edges are removed randomly from the network while ensuring that the rest network structure is still connected. The proximity of all pair of nodes is calculated in the test. AUC is used as the evaluation metric.

From the results in Table 6.7, HHNE outperforms the baselines upon all the dimensionality, especially in the low dimensionality. The results can demonstrate the generalization ability of HHNE. In DBLP dataset, the results of HHNE in 10 dimensionality exceed all the baselines in higher dimensionality results. In MovieLens dataset, HHNE with only 2 dimensionality surpasses baselines in all dimensionality. Besides, both of LINE(1st) and PoincaréEmb preserve proximities of node pairs linked by an edge, while LINE(1st) embed network into Euclidean spaces and Poincaré embed network into hyperbolic spaces. PoincaréEmb performs better than LINE(1st) in most cases, especially in dimensionality lower than 10, suggesting the superiority of embedding network into hyperbolic spaces. Because HHNE

Table 6.6 AUC scores for network reconstruction. The best results are marked by bold numbers

Dataset	Edge	Dimension	Deepwalk	LINE(1st)	LINE(2nd)	metapath2vec	PoincaréEmb	HHNE
DBLP	P-A	2	0.6933	0.5286	0.6740	0.6686	0.8251	**0.9835**
		5	0.8034	0.5397	0.7379	0.8261	0.8769	**0.9838**
		10	0.9324	0.6740	0.7541	0.9202	0.8921	**0.9887**
		15	0.9666	0.7220	0.7868	0.9500	0.8989	**0.9898**
		20	0.9722	0.7457	0.7600	0.9623	0.9024	**0.9913**
		25	0.9794	0.7668	0.7621	0.9690	0.9034	**0.9930**
	P-V	2	0.7324	0.5182	0.6242	0.7286	0.5718	**0.8449**
		5	0.7906	0.5500	0.6349	0.9072	0.5529	**0.9984**
		10	0.8813	0.7070	0.6333	0.9691	0.6271	**0.9985**
		15	0.9353	0.7295	0.6343	0.9840	0.6446	**0.9985**
		20	0.9505	0.7369	0.6444	0.9879	0.6600	**0.9985**
		25	0.9558	0.7436	0.6440	0.9899	0.6760	**0.9985**
MoiveLens	M-A	2	0.6320	0.5424	0.6378	0.6404	0.5231	**0.8832**
		5	0.6763	0.5675	0.7047	0.6578	0.5317	**0.9168**
		10	0.7610	0.6202	0.7739	0.7231	0.5404	**0.9211**
		15	0.8244	0.6593	0.7955	0.7793	0.5479	**0.9221**
		20	0.8666	0.6925	0.8065	0.8189	0.5522	**0.9239**
		25	0.8963	0.7251	0.8123	0.8483	0.5545	**0.9233**
	M-D	2	0.6626	0.5386	0.6016	0.6589	0.6213	**0.9952**
		5	0.7263	0.5839	0.6521	0.7230	0.7266	**0.9968**
		10	0.8246	0.6114	0.6969	0.8063	0.7397	**0.9975**
		15	0.8784	0.6421	0.7112	0.8455	0.7378	**0.9972**
		20	0.9117	0.6748	0.7503	0.8656	0.7423	**0.9982**
		25	0.9345	0.7012	0.7642	0.8800	0.7437	**0.9992**

can preserve high-order network structure and handle different types of nodes in HG, HHNE is more effective than PoincaréEmb.

More detailed introduction of HHNE can be found in [204].

6.5   Conclusion

The study of hyperbolic representation learning has shown very promising results. In this chapter, we have introduced three hyperbolic GNNs, e.g., HAT, LGCN, and HHNE. HAT leverages hyperbolic graph operations to learn the hyperbolic graph representations. Also, LGCN further proposes some hyperbolic graph operations defined in the hyperboloid model to improve the effectiveness of the hyperbolic graph representations. Moreover, HHNE aims to learn the heterogeneous graph representations in hyperbolic spaces, and the related

Table 6.7 AUC scores for link prediction. The best results are marked by bold numbers

Dataset	Edge	Dimension	Deepwalk	LINE(1st)	LINE(2nd)	metapath2vec	PoincaréEmb	HHNE
DBLP	P-A	2	0.5813	0.5090	0.5909	0.6536	0.6742	**0.8777**
		5	0.7370	0.5168	0.6351	0.7294	0.7381	**0.9041**
		10	0.8250	0.5427	0.6510	0.8279	0.7699	**0.9111**
		15	0.8664	0.5631	0.6582	0.8606	0.7743	**0.9111**
		20	0.8807	0.5742	0.6644	0.8740	0.7806	**0.9106**
		25	0.8878	0.5857	0.6782	0.8803	0.7830	**0.9117**
	P-V	2	0.7075	0.5160	0.5121	0.7059	0.8257	**0.9331**
		5	0.7197	0.5663	0.5216	0.8516	0.8878	**0.9409**
		10	0.7292	0.5873	0.5332	0.9248	0.9113	**0.9619**
		15	0.7325	0.5896	0.5425	0.9414	0.9142	**0.9625**
		20	0.7522	0.5891	0.5492	0.9504	0.9185	**0.9620**
		25	0.7640	0.5846	0.5512	0.9536	0.9192	**0.9612**
MoiveLens	M-A	2	0.6278	0.5053	0.5712	0.6168	0.5535	**0.7715**
		5	0.6353	0.5636	0.5874	0.6212	0.5779	**0.8255**
		10	0.6680	0.5914	0.6361	0.6332	0.5984	**0.8312**
		15	0.6791	0.6184	0.6442	0.6382	0.5916	**0.8319**
		20	0.6868	0.6202	0.6596	0.6453	0.5988	**0.8318**
		25	0.6890	0.6256	0.6700	0.6508	0.5995	**0.8309**
	M-D	2	0.6258	0.5139	0.6501	0.6191	0.5856	**0.8520**
		5	0.6482	0.5496	0.6607	0.6332	0.6290	**0.8967**
		10	0.6976	0.5885	0.7499	0.6687	0.6518	**0.8984**
		15	0.7163	0.6647	0.7756	0.6702	0.6715	**0.9007**
		20	0.7324	0.6742	0.7982	0.6746	0.6821	**0.9000**
		25	0.7446	0.6957	0.8051	0.6712	0.6864	**0.9018**

optimization strategies are derived to optimize the hyperbolic representations. Hyperbolic graph representations have shown their powerful ability in modeling graphs. We hope more hyperbolic GNNs with deeper insights can be proposed to in the future.

6.6   Further Reading

In recent years, representation learning in hyperbolic spaces has received increasing attention. There are also some hyperbolic GNNs been proposed. For example, HGCN [21] leverages hyperbolic graph convolution to learn the node representations in hyperbolic spaces. Different from learning node-level representations, HGNN [120] learns graph-level representations in hyperbolic spaces and achieves state-of-the-art results. Also, some efforts begin to design GNNs with more geometries. GIL [256] combines hyperbolic and Euclidean spaces to design graph convolutional networks. $\kappa-$GCN [5] designs a universal GNN in constant curvature spaces, i.e., Euclidean, hyperbolic, and spherical spaces. Moreover, there are also some studies about hyperbolic representation learning in other fields, such as graph embedding, natural language processing, and recommender systems. For graph embedding, PoincaréEmb [141] embeds graph into hyperbolic spaces to learn the hierarchical feature representation. De et al. [161] propose a novel combinatorial embedding approach as well as an approach to multi-dimensional scaling in hyperbolic spaces. LorentzEmb [142] focuses on discovering pairwise hierarchical relations between concepts via embedding graphs in hyperbolic spaces. For natural language processing, inspired by [141], Dhingra et al. [41] embed text in Poincaré model to learn the words and sentences in hyperbolic spaces. HyperQA [183] focuses on question answering problem, modeling the relationship between question and answer representations in hyperbolic spaces. Leimeister et al. [111] model the words in a hyperboloid model to learn the word representations. PoincaréGlove [185] is a novel word embedding method, which learns the word representations in a Cartesian product of hyperbolic spaces. For recommender systems, Vinh et al. [190] design a recommendation algorithm based on Bayesian personalized ranking in hyperbolic spaces. Chamberlain et al. [20] present a large-scale recommender system in hyperbolic spaces by using Einstein midpoint. A comprehensive review of non-Euclidean deep learning can be found in [153].

Distilling Graph Neural Networks

Jiawei Liu

7.1 Introduction

With the success of deep learning, methods based on Graph Neural Networks (GNNs) have become an effective way to address the graph learning problem by converting the graph data into a low-dimensional space while keeping both the structure and feature information. Recently, the rapid evolution of GNNs has led to a growing number of new architectures as well as novel applications [46, 123, 177]. However, there are still some open problems. For example, how to improve the performance of arbitrary GNN, and how to deploy GNN on resource-constrained devices.

To solve the problems mentioned above, Knowledge Distillation (KD) provides a feasible solution. Knowledge distillation was proposed by Hinton et al. [79] for model compression, i.e., to supervise the training of a compact yet efficient student model by capturing and transferring the knowledge from a large complicated teacher model. Besides the motivation of compressing models, a recent study [57] found that a student can even outperform its teacher in prediction performance if they are parameterized identically. In fact, knowledge distillation has been widely used in many scenarios, including model performance improvement, model compression, model interpretability analysis, data privacy protection, etc.

In this chapter, we will introduce some representative works which apply knowledge distillation to graph neural networks. In Sect. 7.2, we will introduce a KD-based framework (named CPF) which combines label propagation and feature transformation to naturally preserve structure-based and feature-based prior knowledge and improves the performance of any GNNs therefore. In Sect. 7.3, we will introduce a framework (named LTD) which aims

J. Liu (✉)
Beijing University of Posts and Telecommunications, Beijing, China

© The Author(s), under exclusive license to Springer Nature Switzerland AG 2023
C. Shi et al., *Advances in Graph Neural Networks*, Synthesis Lectures
on Data Mining and Knowledge Discovery,
https://doi.org/10.1007/978-3-031-16174-2_7

to improve distillation quality for GNN models by learning node-specific temperatures. In Sect. 7.4, we will introduce a data-free adversarial knowledge distillation framework (named DFAD-GNN), which can distill teacher model without training data.

7.2   Prior-Enhanced Knowledge Distillation for GNNs

7.2.1   Overview

Most GNN models adopt a message passing strategy [61]: each node aggregates features from its neighborhood and then a layer-wise projection function with a non-linear activation will be applied to the aggregated information. In this way, GNNs can utilize both graph structure and node feature information in their models. However, the entanglement of graph topology, node features, and projection matrices in GNNs leads to a complicated prediction mechanism and could not take full advantage of prior knowledge lying in the data. Recent studies proposed to incorporate the label propagation mechanism into GCN by adding regularizations [192] or manipulating graph filters [117, 168], and their experimental results show that GCN can be improved by emphasizing such structure-based prior knowledge. Nevertheless, these methods have three major drawbacks: (1) the main bodies of their models are still GNNs and thus hard to fully utilize the prior knowledge; (2) they are single models rather than frameworks, and thus not compatible with other advanced GNN architectures; and (3) they ignored another important prior knowledge, i.e., *feature-based prior*, which means that a node's label is purely determined by its features.

To address these issues, we propose an effective knowledge distillation framework to inject the knowledge of an arbitrary learned GNN (teacher model) into a well-designed student model. The student model is built with two simple prediction mechanisms, i.e., label propagation and feature transformation, which naturally preserves structure-based and feature-based prior knowledge, respectively. Besides, it has been recognized that the knowledge of a teacher model lies in its soft predictions [79]. By simulating the soft labels predicted by a teacher model, our student model is able to further make use of the knowledge in pretrained GNNs. Consequently, the learned student model has a more interpretable prediction process and can utilize both GNN and structure/feature-based priors. Experimental results show that a student model is able to outperform its corresponding teacher model by $1.4 \sim 4.7\%$ in terms of classification accuracy.

7.2.2   The CPF Method

In this section, we will first present our knowledge distillation framework to extract the knowledge of GNNs. Afterward, we will propose the architecture of our student model, which is a trainable combination of parameterized label propagation and feature-based 2-layer MLP. The full architecture of the model is shown in Fig. 7.1.

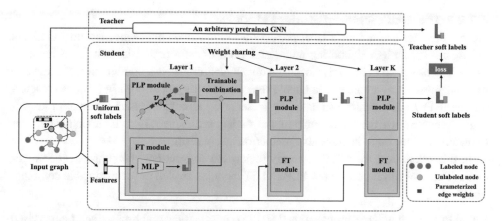

Fig. 7.1 An illustration of the architecture of our proposed student model. Taking the center node v as an example, the student model starts from node v's raw features and a uniform label distribution as soft labels. Then at each layer, the soft label prediction of v will be updated as a trainable combination of Parameterized Label Propagation (PLP) from v's neighbors and Feature Transformation (FT) of v's features. Finally, the distance between the soft label predictions of student and pretrained teacher will be minimized

7.2.2.1 The Knowledge Distillation Framework

Node classification approaches including GNNs can be summarized as a black box that outputs a classifier f given graph structure G, labeled node set V_L and node feature X as inputs. The classifier f will predict the probability $f(v, y)$ that unlabeled node $v \in V_U$ has label $y \in Y$, where $\sum_{y' \in Y} f(v, y') = 1$. For labeled node v, we set $f(v, y) = 1$ if v is annotated with label y and $f(v, y') = 0$ for any other label y'. We use $f(v) \in \mathbb{R}^{|Y|}$ to denote the probability distribution over all labels for brevity.

The teacher model employed in our framework can be an arbitrary GNN model such as GCN [102] or GAT [189]. We denote the pretrained classifier in a teacher model as f_{GNN}. On the other hand, we use $f_{STU;\Theta}$ to denote the student model parameterized by Θ and $f_{STU;\Theta}(v) \in \mathbb{R}^{|Y|}$ represents the predicted probability distribution of node v by the student.

In knowledge distillation [79], the student model is trained to mimic the soft label predictions of a pretrained teacher model. As a result, the knowledge lying in the teacher model will be extracted and injected into the learned student. Therefore, the optimization objective which aligns the outputs between the student model and pretrained teacher model can be formulated as

$$\min_{\Theta} \sum_{v \in V} \text{distance}(f_{GNN}(v), f_{STU;\Theta}(v)), \qquad (7.1)$$

where $\text{distance}(\cdot, \cdot)$ measures the distance between two predicted probability distributions. Specifically, we use Euclidean distance in this work.

7.2.2.2 The Architecture of Student Model

We hypothesize that a node's label prediction follows two simple mechanisms: (1) label propagation from its neighboring nodes and (2) a transformation from its own features. Therefore, as shown in Fig. 7.1, we design our student model as a combination of these two mechanisms, i.e., a Parameterized Label Propagation (PLP) module and a Feature Transformation (FT) module, which can naturally preserve the structure/feature-based prior knowledge, respectively. After the distillation, the student will benefit from both GNN and prior knowledge with a more interpretable prediction mechanism.

In this subsection, we will first briefly review the conventional label propagation algorithm. Then we will introduce our PLP and FT modules as well as their trainable combinations.

Label Propagation Label propagation (LP) [257] is a classical graph-based semi-supervised learning model. This model simply follows the assumption that nodes linked by an edge (or occupying the same manifold) are very likely to share the same label. Based on this hypothesis, labels will propagate from labeled nodes to unlabeled ones for predictions.

Formally, we use f_{LP} to denote the final prediction of LP and f_{LP}^k to denote the prediction of LP after k iterations. In this work, we initialize the prediction of node v as a one-hot label vector if v is a labeled node. Otherwise, we will set a uniform label distribution for each unlabeled node v, which indicates that the probabilities of all classes are the same at the beginning. The initialization can be formalized as

$$f_{LP}^0(v) = \begin{cases} (0, ...1, ...0) \in \mathbb{R}^{|Y|}, & \forall v \in V_L \\ (\frac{1}{|Y|}, ... \frac{1}{|Y|}, ... \frac{1}{|Y|}) \in \mathbb{R}^{|Y|}, & \forall v \in V_U \end{cases} \tag{7.2}$$

where $f_{LP}^k(v)$ is the predicted probability distribution of node v at iteration k. In the $k + 1$th iteration, LP will update the label predictions of each unlabeled node $v \in V_U$ as follows:

$$f_{LP}^{k+1}(v) = (1 - \lambda)\frac{1}{|N_v|} \sum_{u \in N_v} f_{LP}^k(u) + \lambda f_{LP}^k(v), \tag{7.3}$$

where N_v is the set of node v's neighbors in the graph and λ is a hyper-parameter controlling the smoothness of node updates.

Note that LP has no parameters to be trained and thus can not fit the output of a teacher model through end-to-end training. Therefore, we retrofit LP by introducing more parameters to increase its capacity.

Parameterized Label Propagation Module Now, we will introduce our Parameterized Label Propagation (PLP) module by further parameterizing edge weights in LP. As shown in Eq. 7.3, LP model treats all neighbors of a node equally during the propagation. However, we hypothesize that the importance of different neighbors to a node should be different, which determines the propagation intensities between nodes. To be more specific, we assume that the label predictions of some nodes are more "confident" than others: e.g., a node whose predicted label is similar to most of its neighbors. Such nodes will be more likely to propagate their labels to neighbors and keep themselves unchanged.

Formally, we will assign a confidence score $c_v \in \mathbb{R}$ to each node v. During the propagation, all node v's neighbors and v itself will compete to propagate their labels to v. Following the intuition that a larger confidence score will have a larger edge weight, we rewrite the prediction update function in Eq. 7.3 for f_{PLP} as follows:

$$f_{PLP}^{k+1}(v) = \sum_{u \in N_v \cup \{v\}} w_{uv} f_{PLP}^k(u), \tag{7.4}$$

where w_{uv} is the edge weight between node u and v computed by the following softmax function:

$$w_{uv} = \frac{exp(c_u)}{\sum_{u' \in N_v \cup \{v\}} exp(c_{u'})}. \tag{7.5}$$

Similar to LP, $f_{PLP}^0(v)$ is initialized as Eq. 7.2 and $f_{PLP}^k(v)$ remains the one-hot ground truth label vector for every labeled node $v \in V_L$ during the propagation.

Note that we can further parameterize confidence score c_v for inductive setting as an optional choice:

$$c_v = z^T X_v, \tag{7.6}$$

where $z \subset \mathbb{R}^d$ is a learnable parameter that projects node v's feature into the confidence score.

Feature Transformation Module Note that PLP module which propagates labels through edges emphasizes the structure-based prior knowledge. Thus we also introduce Feature Transformation (FT) module as a complementary prediction mechanism. The FT module predicts labels by only looking at the raw features of a node. Formally, denoting the prediction of FT module as f_{FT}, we apply a 2-layer MLP[1] followed by a softmax function to transform the features into soft label predictions:

$$f_{FT}(v) = \text{softmax}(MLP(X_v)). \tag{7.7}$$

A Trainable Combination Now, we will combine the PLP and FT modules as the full model of our student. In detail, we will learn a trainable parameter $\alpha_v \in [0, 1]$ for each node v to balance the predictions between PLP and FT. In other words, the prediction from FT module will be incorporated into that from PLP at each propagation step. We name the full student model as combination of parameterized label propagation and Feature transformation (CPF) and thus the prediction update function for each unlabeled node $v \in V_U$ in Eq. 7.4 will be rewritten as

$$f_{CPF}^{k+1}(v) = \alpha_v \sum_{u \in N_v \cup \{v\}} w_{uv} f_{CPF}^k(u) + (1 - \alpha_v) f_{FT}(v), \tag{7.8}$$

---

[1] We find that 2-layer MLP is necessary for increasing the model capacity of our student, though a single layer logistic regression is more interpretable.

where edge weight w_{uv} and initialization $f^0_{CPF}(v)$ are the same with PLP module. Whether parameterizing confidence score c_v as Eq. 7.6 or not will lead to inductive/transductive variants CPF-ind/CPF-tra.

7.2.2.3 The Overall Algorithm and Details

Assuming that our student model has a total of K layers, the distillation objective in Eq. 7.1 can be detailed as

$$\min_{\Theta} \sum_{v \in V_U} \| f_{GNN}(v) - f^K_{CPF;\Theta}(v) \|_2, \tag{7.9}$$

where $\| \cdot \|_2$ is the L2-norm and the parameter set Θ includes the balancing parameters between PLP and FT $\{\alpha_v, \forall v \in V\}$, confidence parameters in PLP module $\{c_v, \forall v \in V\}$ (or parameter z for inductive setting), and the parameters of MLP in FT module Θ_{MLP}. There is also an important hyper-parameter in the distillation framework: the number of propagation layers K.

7.2.3 Experiments

7.2.3.1 Experimental Settings

Datasets We use five public benchmark datasets (i.e. Cora, Citeseer, Pubmed, A-Computers, and A-Photo) for experiments. As previous works [104, 165, 175] did, we only consider the largest connected component and regard the edges as undirected. Following the experimental settings in previous work [165], we randomly sample 20 nodes from each class as labeled nodes, 30 nodes for validation, and all other nodes for test.

Model Settings For a thorough comparison, we consider seven GNN models as teacher models in our knowledge distillation framework, that is, GCN [102], GAT [189], APPNP [103], SAGE [74], SGC [208], GCNII [23], and GLP [117].

For each dataset and teacher model, we test the following student variants:

- PLP: The student variant with only the Parameterized Label Propagation mechanism;
- FT: The student variant with only the Feature Transformation mechanism;
- CPF-ind: The full model CPF with inductive setting;
- CPF-tra: The full model CPF with transductive setting.

7.2.3.2 Analysis of Classification Results

Experimental results on five datasets with two GNN teachers and four student variants are presented in Tables 7.1 and 7.2. We find that the proposed knowledge distillation framework accompanying the full architecture of student model CPF-ind and CPF-tra is able to improve the performance of the corresponding teacher model consistently and significantly.

Table 7.1 Classification accuracies with teacher models as GCN [102]

Datasets	Teacher	Student variants				+Impv. (%)
	GCN	PLP	FT	CPF-ind	CPF-tra	
Cora	0.8244	0.7522	0.8253	**0.8576**	0.8567	4.0
Citeseer	0.7110	0.6602	0.7055	0.7619	**0.7652**	7.6
Pubmed	0.7804	0.6471	0.7964	0.8080	**0.8104**	3.8
A-Computers	0.8318	0.7584	0.8356	**0.8443**	**0.8443**	1.5
A-Photo	0.9072	0.8499	0.9265	**0.9317**	0.9248	2.7

Table 7.2 Classification accuracies with teacher models as GAT [189]

Datasets	Teacher	Student variants				+Impv. (%)
	GAT	PLP	FT	CPF-ind	CPF-tra	
Cora	0.8389	0.7578	0.8426	0.8576	**0.8590**	2.4
Citeseer	0.7276	0.6624	0.7591	0.7657	**0.7691**	5.7
Pubmed	0.7702	0.6848	0.7896	0.8011	**0.8040**	4.4
A-Computers	0.8107	0.7605	0.8135	**0.8190**	0.8148	1.0
A-Photo	0.8987	0.8496	0.9190	**0.9221**	0.9199	2.6

The more detailed method description and experiment validation can be seen in [219].

7.3   Temperature-Adaptive Knowledge Distillation for GNNs

7.3.1   Overview

Recently, there is an emerging trend that equips GNNs with knowledge distillation for better efficiency or effectiveness. Besides the choice of teacher and student, the distillation process, which determines how the soft predictions of teacher and student models are aligned, is also vital to the prediction performance of a distilled student on downstream tasks [79]. However, to the best of our knowledge, existing knowledge distillation methods applied on GNNs all employed predefined distillation processes, i.e., with only hyper-parameters but without any learnable parameters. In other words, the distillation processes are designed heuristically or empirically without any supervision from the performance of distilled students, which isolates distillation from evaluation and thus leads to suboptimal results.

In this work, we aim to propose a general knowledge distillation framework that can be applied to any pretrained GNN models to further improve their performance. To overcome the isolation problem between distillation and evaluation, instead of introducing the global temperature as a hyper-parameter, we innovatively propose to learn node-specific temperatures supervised by the performance of distilled student models, as shown in Fig. 7.2.

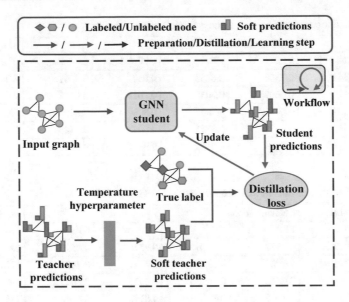

(a) The typical knowledge distillation framework applied on GNNs.

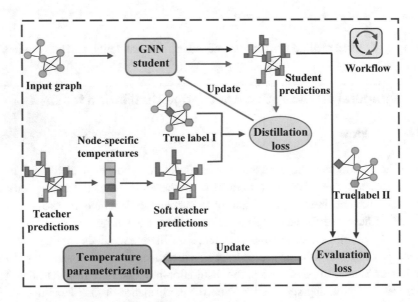

(b) Our proposed knowledge distillation framework.

Fig. 7.2 An illustration of **a** the typical framework [79] for distilling GNNs [218–220]; and **b** our proposed distillation framework. Instead of introducing a unified temperature hyper-parameter, we will learn node-specific temperatures supervised by the performance of distilled GNN student based on a novel iterative workflow. Note that the true labels used in the typical framework is divided to two subsets in our framework, and thus we did not employ additional training data

Specifically, we parameterize each node's temperature by a function of its neighborhood's encodings and predictions. Due to the isolation problem in traditional knowledge distillation frameworks [79], the partial derivative of a distilled student's performance with respect to node temperatures does not exist, which makes it non-trivial to learn the parameters in temperature parameterization. Therefore, we design a novel iterative learning process, which alternatively performs distillation steps and learning steps for parameter training. Experimental results show that on average our distilled GNN student has more than 2% relative improvements in prediction accuracy, respectively.

7.3.2 The LTD Method

7.3.2.1 Knowledge Distillation On Graph Neural Networks

We simply formalize a GNN encoder in a black box form as

$$f_{v;\Theta} = \text{GNN}(v|G, X; \Theta) \in \mathbb{R}^{|Y|}, \; p_{v;\Theta} = \text{softmax}(f_{v;\Theta}), \tag{7.10}$$

where Θ is the learnable parameters in the GNN and $p_{v;\Theta}$ is the predicted label distribution normalized by the softmax operator.

Then GNNs will minimize the distance between ground truth label y_v and predicted label $p_{v;\Theta}$ for each labeled node $v \in V_L$, and usually employ the cross-entropy loss to train the parameters Θ:

$$\min_{\Theta} \sum_{v \in V_L} \mathcal{L}_{CE}(y_v, p_{v;\Theta}), \tag{7.11}$$

$$\mathcal{L}_{CE}(y_v, p_{v;\Theta}) = -\sum_{i=1}^{|Y|} y_v[i] \cdot \log p_{v;\Theta}[i], \tag{7.12}$$

where $y_v[i]$ and $p_{v;\Theta}[i]$ are, respectively, the ith entry of vector y_v and $p_{v;\Theta}$.

In this work, we simply let the teacher and student models have the same neural architecture as suggested by BAN [57], and denote them as GNN_T and GNN_S with parameters Θ_T and Θ_S, respectively. Given pretrained parameters Θ_T of the teacher model GNN_T (learned by Eq. (7.11)), we will train the parameters Θ_S of the student model GNN_S by aligning the soft predictions between GNN_T and GNN_S.

Formally, the knowledge distillation framework aims at optimizing

$$\min_{\Theta_S} \sum_{v \in V} \mathcal{L}_{CE}(p_{v;\Theta_T}, p_{v;\Theta_S}) + \lambda \sum_{v \in V_L} \mathcal{L}_{CE}(y_v, p_{v;\Theta_S}), \tag{7.13}$$

where the first term is the cross-entropy with the teacher's predictions, the second term is the cross-entropy with ground truth labels on V_L, and λ is the balance hyper-parameter.

Note that many knowledge distillation methods since [79] will soften both teacher and student's predictions $p_{v;\Theta_T}, p_{v;\Theta_S}$ in the first term of Eq. (7.13) before distillation, by introducing extra temperature hyper-parameters:

$$p_{v;\Theta_T}(\tau_v^T) = \text{softmax}(f_{v;\Theta_T}/\tau_v^T),$$
$$p_{v;\Theta_S}(\tau_v^S) = \text{softmax}(f_{v;\Theta_S}/\tau_v^S), \quad (7.14)$$

where $\tau_v^T, \tau_v^S \in \mathbb{R}_+$ are temperature hyper-parameters. A temperature of 1 corresponds to the original softmax operation. Larger temperatures will produce softer predictions (toward uniform distribution), while smaller temperatures will produce harder predictions (toward one-hot distribution). In the most popular distillation framework [79], all the temperatures are set as the same hyper-parameter τ, i.e., $\tau_v^T = \tau_v^S = \tau$ for every node v. By tuning the global temperature hyper-parameter, the distilled GNN student is then evaluated and expected to have better performance than the teacher.

7.3.2.2 Learning Node-specific Temperatures for Better Distillation

We will first present how we introduce learnable parameters in temperature parameterization, and then design a novel algorithm for parameter training based on an iterative learning process.

Temperature Parameterization Directly assigning each node a free parameter as node-specific temperature would lead to a serious overfitting problem. Therefore, we assume that each node v's temperature can be parameterized by a function of the student's logit vector of v and the teacher's predicted distributions of v's neighbors. In this way, nodes with similar encodings and neighborhood predictions will have similar temperatures.

Specifically, we set all the student temperatures τ_v^S to 1 for more calibrated predictions [246], and parameterize teacher temperature τ_v^T as

$$\tau_{v;\Theta_S,\Theta_T,\Omega}^T = \text{MLP}(\text{Concat}(f_{v;\Theta_S}, ||f_{v;\Theta_S}||_2, e_{v;\Theta_T}); \Omega), \quad (7.15)$$

where $\text{MLP}(\cdot; \Omega)$ denotes a multi-layer perception with parameters Ω, Concat() is the concatenation operator, $|| \cdot ||_2$ is the L2-norm, and $e_{v;\Theta_T}$ is defined as the entropy of the average predictions of v's neighbors:

$$e_{v;\Theta_T} = \mathcal{L}_{CE}(\frac{1}{|N_v|} \sum_{u \in N_v} p_{u;\Theta_T}, \frac{1}{|N_v|} \sum_{u \in N_v} p_{u;\Theta_T}), \quad (7.16)$$

where N_v is the set of v's neighbors.

Here, $f_{v;\Theta_S}$ and $||f_{v;\Theta_S}||_2$ depend on the student parameter Θ_S, while $e_{v;\Theta_T}$ is a node-specific constant since the teacher parameter Θ_T is pretrained and fixed. We will investigate the effect of each concatenated component $f_{v;\Theta_S}, ||f_{v;\Theta_S}||_2, e_{v;\Theta_T}$, and discuss the learned temperatures in our experiments. In addition, to avoid the gradient explosion or vanishment issue, we also restrict the temperatures within range $[l, r]$ by a function based on sigmoid operation $(r - l)\sigma(\cdot) + l$.

Iterative Learning Process In order to supervise the training of node-specific temperatures, we partition the labeled node set V_L into two disjoint sets V_{Dis} and V_{Temp}: V_{Dis} is still used in the second term of Eq. (7.13) for distillation, while V_{Temp} is used for evaluating distilled students and learning temperatures.

Formally, the loss for the distillation part can be written as

$$\mathcal{L}_{Dis}(\Theta_S, \Omega) = \sum_{v \in V} \mathcal{L}_{CE}(p_{v;\Theta_T}(\tau^T_{v;\Theta_S,\Theta_T,\Omega}), p_{v;\Theta_S})$$
$$+ \lambda \sum_{v \in V_{Dis}} \mathcal{L}_{CE}(y_v, p_{v;\Theta_S}), \qquad (7.17)$$

and the loss for evaluating distilled students and supervising temperatures is

$$\mathcal{L}_{Temp}(\Theta_S) = \sum_{v \in V_{Temp}} \mathcal{L}_{CE}(y_v, p_{v;\Theta_S}). \qquad (7.18)$$

However, due to the isolation between distillation and evaluation, the evaluation loss $\mathcal{L}_{Temp}$ is only related to the parameters Θ_S of student model and the partial derivative $\partial \mathcal{L}_{Temp}/\partial \Omega$ does not exist, which makes it impossible to learn the temperatures via back-propagation.

To address this problem, we propose a novel iterative learning process by alternatively performing the following distillation and learning steps:

Distillation step. According to the distillation loss in Eq. (7.17), we update the parameters Θ_S through back-propagation for a single step:

$$\Theta'_S := \Theta_S - \alpha \frac{\partial \mathcal{L}_{Dis}(\Theta_S, \Omega)}{\partial \Theta_S}, \qquad (7.19)$$

where α is the learning rate for distillation step.

Learning step. We evaluate $\mathcal{L}_{Temp}$ with the updated parameter Θ'_S, and then perform back-propagation on Ω by the chain rule:

$$\Omega' := \Omega - \beta \frac{\partial \mathcal{L}_{Temp}(\Theta'_S)}{\partial \Theta'_S} \frac{\partial \Theta'_S}{\partial \Omega}, \qquad (7.20)$$

where β is the learning rate for learning step.

Here, we decompose the partial derivative of $\mathcal{L}_{Temp}$ with respect to Ω into the product of $\partial \mathcal{L}_{Temp}(\Theta'_S)/\partial \Theta'_S$ and $\partial \Theta'_S/\partial \Omega$, which can be calculated by the partial derivative of Eqs. (7.18) and (7.19), respectively. By iteratively executing the distillation and learning steps, we can learn node-specific temperatures parameterized by Ω toward better prediction performance of distilled students.

Implementation Details We name our proposed framework as LTD (Learning To Distill). Depending on the teacher GNN model, our distilled student model can be correspondingly named LTD-GCN, LTD-GAT, etc.

We bisect the labeled node set V_L into V_{Dis} and V_{Temp}, i.e., $|V_{Dis}| = |V_{Temp}|$. We will run 20 epochs of distillation without updating Ω as warmup, and then perform the iterative learning process. The time complexity of each iteration in LTD is linear with respect to the number of nodes and edges. In fact, a single run of LTD can be finished in 5 minutes with a single GPU device of GeForce GTX 1080 Ti for all datasets.

7.3.3 Experiments

7.3.3.1 Experimental Setup

Datasets In our experiments, we employ five benchmark datasets (i.e. Cora, Citeseer, Pubmed, A-Computers, and A-Photo) widely used in former works [165, 219]. Specifically, we use the largest connected component as previous works did [165, 219].

Teacher/Student Models In our experiments, we use two representative GNN models as teacher/student models, i.e., GCN [102] and SAGE [74]. We use a two-layer setting for all the models.

Framework Variants For each dataset, we test the following frameworks based on different GNN teacher/student models:

- FT (Fixed Temperature): All nodes use the same temperature as a hyper-parameter, which is adopted in most knowledge distillation frameworks.
- LTD$_{w/o \, LS}$: The proposed LTD method without learning steps, i.e., the parameter Ω is never updated after initialization.
- LTD: The proposed method.

Experimental Settings We conduct experiments on the most popular task for evaluating GNNs, i.e., semi-supervised node classification. For each dataset, we use 40 nodes per class as the training data, 10 nodes per class as the validation data, and all the rest nodes for testing. For each combination of GNN model and dataset, we will pretrain a GNN model as the teacher and fix its parameters. After the distillation, we will evaluate the distilled students learned by different frameworks. For evaluation metric, we will report the classification accuracy as previous work [102, 189] did.

For the hyper-parameter searching of LTD, we conduct heuristic search by exploring the learning rate of distillation step $\alpha \in [1e - 6, 1e - 3]$, the learning rate of learning step $\beta \in [1e - 6, 1e - 1]$ and the balance hyper-parameter λ from $\{0.1, 1, 50, 100, 200\}$ with the help of Optuna,[2] an automatic hyper-parameter optimization toolkit. For the framework variant FT, we conduct a careful grid search of global temperature τ from $\{0.001, 0.01, 0.1, 1, 4, 8, 12, 16, 20, 24\}$ and balance hyper-parameter λ from $\{0.1, 1, 50, 100, 200\}$, and employ Adam optimizer with learning rate 0.01 for updating parameters.[3]

---

[2] https://optuna.org/.

[3] We find that Adam optimizer gives better performance than our gradient descent strategy (i.e., a fixed learning rate $\alpha \in [1e - 6, 1e - 3]$) for FT.

For other settings, we always use a 2-layer MLP in temperature parameterization with hidden dimension as 64 and dropout rate as 0.6. The ranges of temperatures are within $[-1, 4]$. For all methods, we use early stopping with a patience of 50.

7.3.3.2 Analysis of Classification Results

We present the results on five benchmark datasets with two GNN models in Tables 7.3 and 7.4. We bold the best results among the teacher model and the three distilled students learned by different framework variants. The relative improvements over the better model between teacher and FT are also reported. The GNN students distilled by our proposed LTD framework can achieve consistent improvements over all two GNN models on the five datasets. The more detailed method descriptions and experiments can be seen in [29].

Table 7.3 Classification accuracies with GNN model as GCN

Dataset	GNN	Framework variants			+Impv. (%)
	GCN	FT	LTD$_{w/o LS}$	LTD	
Citeseer	0.7359	0.7547	0.7586	**0.7851**	3.49
Cora	0.8534	0.8600	0.8614	**0.8721**	1.24
Pubmed	0.7989	0.8029	0.7897	**0.8191**	2.02
A-Computers	0.8594	0.8468	0.8443	**0.8645**	0.59
A-Photo	0.9223	0.9231	0.9032	**0.9324**	1.01

Table 7.4 Classification accuracies with GNN model as GAT

Dataset	GNN	Framework variants			+Impv. (%)
	GAT	FT	LTD$_{w/o LS}$	LTD	
Citeseer	0.7525	0.7564	0.7044	**0.7735**	2.26
Cora	0.8520	0.8534	0.8356	**0.8656**	1.43
Pubmed	0.7944	0.8048	0.8144	**0.8274**	1.60
A-Computers	0.8091	0.8079	0.7823	**0.8304**	2.63
A-Photo	0.9094	0.9194	0.9145	**0.9316**	1.33

7.4   Data-Free Adversarial Knowledge Distillation for GNNs

7.4.1   Overview

Training the powerful GNNs often require heavy computation and storage; hence, it is hard to deploy them into resource-constrained devices. Knowledge distillation (KD) [79] is one of the most popular paradigms for learning a portable student model from the pretrained complicated teacher by directly imitating its outputs. However, the original training data of knowledge distillation is often unavailable due to privacy concerns. An effective way to avert the above-mentioned issue is using synthetic graphs, i.e., data-free knowledge distillation [122, 135], where the data is reversely generated from the pre-trained models. Data-free distillation has received a lot of research in the field of computer vision [48, 122], which is however rarely been explored in graph mining. Note that Deng et al. [39] have made some pilot studies on this problem and proposed graph-free knowledge distillation (GFKD). Unfortunately, GFKD is not an end-to-end approach. It only takes the fixed teacher model into account, ignoring the information from the student model when generating graphs. Thus, these generated graphs are not very useful for distilling the student model efficiently which leads to unsatisfactory performance.

In this work, we propose a novel data-free adversarial knowledge distillation framework for GNNs (DFAD-GNN). DFAD-GNN used a knowledge distillation method based on GAN [63]. DFAD-GNN contains one generator and two discriminators. One fixed discriminator is the pretrained teacher model, and the other is the compact student model that we aim to learn. The generator generates graphs to help transfer teachers' knowledge to students. Unlike previous work [39], our generator can fully utilize both the intrinsic statistics from the pretrained teacher model and the customizable information from the student model, which can help generate high-quality and diverse training data to improve the student model's generalizability.

7.4.2   The DFAD-GNN Method

As shown in Fig. 7.3, DFAD-GNN mainly consists of three components: one generator and two discriminators. One fixed discriminator is the pretrained teacher model $\mathcal{T}$, the other is the compact student model $\mathcal{S}$ that we aim to learn. More specifically, the generator $\mathcal{G}$ takes samples z from a prior distribution and generates fake graphs. Then the generated graphs are used to train a student model under the supervision of the teacher model.

7.4.2.1 Generator
The generator $\mathcal{G}$ is used to synthesize fake graphs that maximize the disagreement between the teacher $\mathcal{T}$ and the student $\mathcal{S}$. $\mathcal{G}$ takes D-dimensional vectors $z \in \mathbb{R}^D$ sampled from

a standard normal distribution $z \sim \mathcal{N}(0, I)$ and outputs graphs. For each z, G outputs an object: $F \in \mathbb{R}^{N \times T}$ that defines node features, here N is the node number and T is node feature dimension. Then we calculate adjacency matrix A as follows:

$$A = \sigma \left(F F^{\top} \right), \tag{7.21}$$

where $\sigma(\cdot)$ is the logistic sigmoid function.

The loss function used for G is the same as that used for S, except that the goal is to maximize it. We formulate this problem as an adversarial game in which G and S compete to respectively maximize and minimize the same function. In other words, the student is trained to match the teacher's predictions and the generator is trained to generate difficult graphs for the student. The adversarial game can be written as

$$\max_{G} \min_{S} \mathbb{E}_{z \sim \mathcal{N}(0,1)}[\mathcal{D}(\mathcal{T}(G(z)), S(G(z)))], \tag{7.22}$$

where $\mathcal{D}(\cdot)$ indicates the differences discrepancy between the teacher $\mathcal{T}$ and the student S.

If generator keeps generating simple and duplicate graphs, student model will fit these graphs, resulting in a very low model discrepancy between student model and teacher model. In this case, the generator is forced to generate difficult and different graphs to enlarge the discrepancy.

7.4.2.2 Adversarial Distillation

As aforementioned, the generator $G(z, \theta^g)$ is used to generate graphs. The student model $S(x, \theta^s)$, together with the teacher model $\mathcal{T}(x, \theta^t)$ are jointly viewed as the discriminator to measure the discrepancy $\mathcal{D}(\mathcal{T}, S; G)$. The adversarial training process consists of two stages: the distillation stage that minimizes the discrepancy; and the generation stage that maximizes the discrepancy, as shown in Fig. 7.3.

Distillation Stage In this stage, we fix the generator G and only update the student S in the discriminator. We sample a batch of random noises z from a standard normal distribution and construct fake graphs with generator G. Then each graph X is fed to both the teacher and the student model to produce the output q^t and q^s, where q is a vector indicating the scores of different categories.

The choice of loss is of paramount importance to a successful distillation. In our approach, the choice of loss involves similar factors to those outlined in research on GANs: multiple works have discussed the problem of vanishing gradients as the discriminator becomes strong in the case of GAN training [4, 72].

Actually, there are several ways to define the model discrepancy to drive the student learning, such as KullbackLeibler Divergence (KLD) and Mean Square Error (MSE). These loss functions are very effective in data-driven KD when the training data is available, yet problematic if directly applied to our framework. An important reason is that, when the

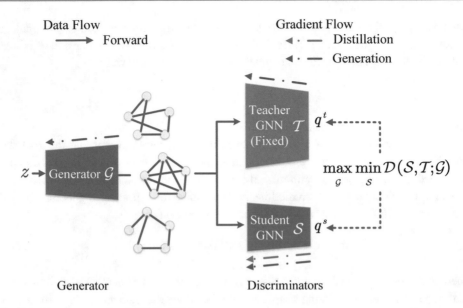

Fig. 7.3 The framework of DFAD-GNN

student converges on the generated graphs, these two loss functions will produce decayed gradients [100], which will deactivate the learning of generator, resulting in a dying min-max game. For our approach, we minimize the Mean Absolute Error (MAE) between q^t and q^s, which provides stable gradients for the generator so that the vanishing gradients can be alleviated. In our experiment, we empirically find that this significantly improves student's performance over other possible losses.

Now we can define the loss function for distillation stage as follows:

$$\mathcal{L}_{DIS} = \mathcal{D}(\mathcal{T}, \mathcal{S}; \mathcal{G}) = \mathbb{E}_{z \sim p_z(z)} \left[\frac{1}{n} \|\mathcal{T}(\mathcal{G}(z)) - \mathcal{S}(\mathcal{G}(z))\|_1 \right]. \tag{7.23}$$

Intuitively, this stage is very similar to KD, but the goals are slightly different. In KD, students can greedily learn from the soft targets produced by the teacher, as these targets are obtained from the real data [79] and contain useful knowledge for the specific task. However, in our setting, we have no access to any real data. The fake graphs synthesized by the generator are not guaranteed to be useful, especially at the beginning of training. As aforementioned, the generator is required to produce graphs to measure the model discrepancy between the teacher model and the student model. Another essential purpose of the distillation stage is to construct a better search space to force the generator to find new graphs.

Generation Stage The goal of the generation stage is to push the generation of new graphs. In this stage, we fix the two discriminators and only update the generator. We encourage the generator to produce more confusing training graphs. A straightforward way to achieve this goal is to take the negative MAE loss as the objective for optimizing the generator:

$$\mathcal{L}_{GEN} = -\mathcal{L}_{DIS} = -\mathbb{E}_{z \sim p_z(z)} \left[\frac{1}{n} \| \mathcal{T}(\mathcal{G}(z)) - \mathcal{S}(\mathcal{G}(z)) \|_1 \right]. \qquad (7.24)$$

With the generation loss, the error first back-propagates through the discriminator, i.e., teacher and student models, then through the generator, yielding the gradients for optimizing the generator.

7.4.3   Experiments

7.4.3.1 Datasets
We adopt three bioinformatics graph classification benchmark datasets including MUTAG, PTC_MR, and PROTEINS. The statistics of these datasets are summarized in Table 7.5. To remove the unwanted bias toward the training data, for all experiments on these datasets, we evaluate the model performance with a ten-fold cross-validation setting, where the dataset split is based on the conventionally used training/test splits [143, 215, 235] with LIBSVM [22]. We report the average and standard deviation of validation accuracies across the ten folds within the cross-validation.

7.4.3.2 Generator Architecture
We adopt a generator with fixed architecture for all experiments. The generator takes a 32-dimensional vector sampled from a standard normal distribution $z \sim \mathcal{N}(\mathbf{0}, \mathbf{I})$. We process it with a three-layer MLP of [64, 128, 256] hidden units respectively, with tanh as activation functions. Eventually, the last layer is linearly projected to match node feature dimensions $\mathbf{F} \in \mathbb{R}^{N \times T}$ and normalized in their last dimension with a softmax operation.

7.4.3.3 Teacher/Student Architecture
To demonstrate the effectiveness of our proposed framework, we consider four GNN models as teacher and student models for a thorough comparison, including: GIN [215], GCN [102], GAT [189], and GraphSAGE [74]. In order to be more unified, we use 5 layers with

Table 7.5 Summary of datasets

Dataset	#Graphs	#Classes	Avg#Graph size
MUTAG	188	2	17.93
PTC_MR	344	2	14.29
PROTEINS	1113	2	39.06

128 hidden units for teacher models. For the student model, we conducted experiments to gradually reduce the number of layers $l \in \{5, 3, 2, 1\}$ and gradually reduce the number of hidden units $h \in \{128, 64, 32, 16\}$. We use a graph classifier layer which first builds a graph representation by averaging all node features extracted from the last GNN layer and then passing this graph representation to an MLP.

7.4.3.4 Baselines

We compare the following baselines to demonstrate the effectiveness of our proposed framework.

- Teacher: The given pretrained model serves as the teacher in the distillation process.
- KD: The generator is removed, and the student model is trained on 100% original training data in our framework.
- RANDOM: The generator's parameters are not updated and the student model is trained on the noisy graphs generated by the randomly initialized generator.
- GFKD: GFKD is a data-free KD for GNNs by modeling the topology of graph with a multinomial distribution [39]. It first learns the fake graphs that the knowledge in the teacher GNN is more likely to concentrate on and then uses these fake graphs to transfer knowledge to the student.

7.4.3.5 Experimental Results

We have pretrained all datasets on GCN, GIN, GAT, and GraphSAGE with 5 layers and 128 hidden units (5–128 for short), and found that GIN performs best on all datasets. Therefore, we adopt GIN as the teacher model in Table 7.6. We choose two representative architectures 1–128 and 5–32 for four kinds of student models. In order to compare the performance of our model across different datasets, we calculate the accuracy percentage of the student model with respect to the teacher model. Besides, we calculate the ratio between the student model's parameters and the teacher model's parameters, i.e., compression ratio.

From Table 7.6, it can be observed that KD's performance is very close to even outperforms the teacher model. That's because KD is a data-driven method which uses the same training data as the teacher model for knowledge distillation. This also implies that the loss function of our DFAD-GNN is very effective in distilling knowledge from the teacher model to the student model, as we apply the same loss function in both KD and our DFAD-GNN.

We also observe that RANDOM delivers the worst performance as the generator is not updated during the training process; thus, the generator will not generate difficult graphs as the student model progresses. Consequently, the student model did not learn enough knowledge from the teacher, resulting in poor results.

In terms of the efficacy of our DFAD-GNN, Table 7.6 shows that our DFAD-GNN consistently outperforms the recent data-free method GFKD [39]. Unlike GFKD that models

Table 7.6 Test accuracies (%) on three datasets. GIN-5–128 means 5 layers GIN with 128 hidden units. $(6.7\% \times m)$ under student model means the percentage of student model parameters to teacher model parameters, m is the number of teacher model parameters. $(90.8\% \times t)$ under DFAD-GNN means the percentage of student model accuracy to teacher model accuracy, t is the accuracy of the corresponding teacher network

Datasets	MUTAG		PTC_MR		PROTEINS	
Teacher	GIN-5-128		GIN-5-128		GIN-5-128	
	96.7±3.7		75.0±3.5		78.3±2.9	
Student	GIN-5-32	GIN-1-128	GIN-5-32	GIN-1-128	GIN-5-32	GIN-1-128
	$(6.7\% \times m)$	$(20.6\% \times m)$	$(6.7\% \times m)$	$(20.6\% \times m)$	$(6.7\% \times m)$	$(20.6\% \times m)$
KD	96.7±5.1	95.3±4.6	76.6±5.9	77.0±8.1	76.0±5.1	78.8±3.2
RANDOM	67.9±8.0	62.9±8.5	60.1±9.1	61.0±8.5	60.8±9.4	60.2±9.2
GFKD	77.8±11.1	72.6±10.4	65.2±7.7	62.1±7.0	61.3±4.0	62.5±3.6
DFAD-GNN	**87.8±6.9**	**85.6±6.7**	**71.0±3.1**	**69.7±3.5**	**70.0±4.2**	**69.9±5.3**
	$(90.8\% \times t)$	$(88.5\% \times t)$	$(94.7\% \times t)$	$(92.9\% \times t)$	$(89.4\% \times t)$	$(89.3\% \times t)$
Student	GCN-5-32	GCN-1-128	GCN-5-32	GCN-1-128	GCN-5-32	GCN-1-128
	$(3.3\% \times m)$	$(10.6\% \times m)$	$(3.3\% \times m)$	$(10.6\% \times m)$	$(3.3\% \times m)$	$(10.6\% \times m)$
KD	86.7±9.4	82.2±10.2	70.9±7.3	70.0±6.9	74.5±3.8	75.5±4.0
RANDOM	58.9±19.3	55.6±21.1	59.4±10.1	55.6±8.1	59.2±8.4	57.9±8.0
GFKD	70.0±11.2	69.1±10.3	65.0±8.2	61.9±8.5	62.9±7.7	61.4±8.8
DFAD-GNN	**74.1±9.3**	**76.4±8.8**	**67.7±2.9**	**67.9±3.5**	**67.2±5.0**	**65.7±3.7**
	$(76.2\% \times t)$	$(79.0\% \times t)$	$(90.3\% \times t)$	$(90.5\% \times t)$	$(85.8\% \times t)$	$(83.9\% \times t)$
Student	GAT-5-32	GAT-1-128	GAT-5-32	GAT-1-128	GAT-5-32	GAT-1-128
	$(164.6\% \times m)$	$(84.5\% \times m)$	$(164.6\% \times m)$	$(84.5\% \times m)$	$(164.6\% \times m)$	$(84.5\% \times m)$
KD	87.8±8.9	82.2±10.2	73.2±5.5	69.7±6.8	76.6±3.4	74.5±4.6
RANDOM	63.9±17.3	57.5±20.3	60.0±7.1	59.4±6.7	59.8±6.4	60.6±5.6
GFKD	72.5±13.8	70.4±11.9	63.2±6.5	62.7±7.0	62.2±6.8	62.8±7.9
DFAD-GNN	**76.9±6.9**	**77.3±5.9**	**66.4±3.9**	**68.0±4.7**	**67.8±4.9**	**66.0±4.7**
	$(79.5\% \times t)$	$(79.9\% \times t)$	$(88.5\% \times t)$	$(90.7\% \times t)$	$(86.6\% \times t)$	$(84.3\% \times t)$
Student	GraphSAGE -5-32	GraphSAGE -1-128	GraphSAGE -5-32	GraphSAGE -1-128	GraphSAGE -5-32	GraphSAGE -1-128
	$(5.9\% \times m)$	$(11.1\% \times m)$	$(5.9\% \times m)$	$(11.1\% \times m)$	$(5.9\% \times m)$	$(11.1\% \times m)$
KD	87.8±12.1	82.8±9.8	75.6±5.3	70.3±6.6	76.3±3.5	75.7±4.5
RANDOM	62.2±17.4	57.8±22.7	61.1±7.0	59.9±6.9	57.4±8.5	55.7±6.3
GFKD	67.7±12.9	68.1±12.1	62.5±5.9	63.0±6.6	63.3±7.7	61.8±7.9
DFAD-GNN	**76.5±7.3**	**75.9±6.5**	**66.9±3.7**	**67.5±3.9**	**69.0±6.1**	**67.8±5.4**
	$(79.1\% \times t)$	$(78.5\% \times t)$	$(89.2\% \times t)$	$(90.0\% \times t)$	$(88.1\% \times t)$	$(86.6\% \times t)$

the graph structures with a multinomial distribution which needs a complicated gradient estimator to optimize, our DFAD-GNN only performs a simple MLP at the generator and uses MAE loss in an end-to-end training process. We conjecture the potential reason that

DFAD-GNN can significantly outperform GFKD is the teacher encodes the distribution characteristics of the original input graphs under its own feature space. Simply inverting graphs in GFKD tends to overfit the partial distribution information stored in this teacher model. As a consequence, their generated fake graphs are lacking generalizability and diversity. In contrast, our generated graphs are more conducive to transferring the knowledge of the teacher model to the student model.

In terms of stability, it can be seen from Table 7.6 that the standard deviation of our DFAD-GNN is the smallest among all the data-free baselines and across all datasets, indicating that our model can obtain relatively stable prediction results.

For the student model, we find that the most expressive architecture is GIN. Because GIN can identify some graph structures that cannot be distinguished by GCN, GAT, and GraphSAGE, such as an isomorphic graph.

Another interesting observation is that the performance of the compressed model is not necessarily worse than the more complex model. As can be seen from Table 7.6 that the performance of a more compressed student model with 5–32 is not necessarily worse than the student model with 1–128. Therefore, we speculate that the performance of the student model may have no obvious relationship with the degree of model compression, which requires further investigation.

7.5   Conclusion

Improving arbitrary GNN models is always a very challenging task, both from the motivation of improving the effectiveness or protecting data privacy. In this chapter, we introduce three knowledge distillation (KD) frameworks to address these challenges, which focus on student models, temperature adaptation, and data deficiency of knowledge distillation, respectively. To be more specific, we first introduce a prior-enhanced KD framework (i.e., CPF), which enhances an arbitrary GNN model by specifically designing a student model with prior knowledge enhancement. Next, we introduce a temperature-adaptive KD framework (i.e., LTD), which improves distillation by learning distillation temperature. Finally, we introduce a data-free adversarial KD framework (i.e., DFAD-GNN), which enables us to distill without training data for the teacher model. Experiments verify the effectiveness of these methods on various GNN models.

As for future work, one can explore to distill graph neural networks on other tasks (e.g., clustering and link prediction), and use knowledge distillation to analyze the interpretability of graph neural networks.

7.6   Further Reading

The research area of distilling graph neural networks is still fast evolving. For a more comprehensive understanding of knowledge distillation, you can read [66], which provides an overview of knowledge distillation. If you would like to learn more about applications of knowledge distillation to improve the efficiency or performance of arbitrary GNNs, you can read the following papers:

For the goal of efficiency, [220] proposed a local structure preserving module to distill the embedded topological structure from a deep GCN to a shallow one. Yan et al. [218] designed a peer-aware module to help shallow student models explore the rich structural information hidden behind higher-order aggregation during distillation. Jing et al. [96] designed a gradient-based topological semantics alignment loss and a slimmable graph convolution layer to support distillation from diversified teachers. Inspired by [219, 238] uses distilled MLP to avoid graph dependency and achieve inference acceleration.

For the goal of performance, [240] considered both node reliability and edge reliability to filter out unreliable information when distilling. Chen et al. [26] proposed a multi-level self-distillation framework to retain high discrepancy of consecutive layers to alleviate over-smoothing. Zhang et al. [239] designed a reception-aware decoupled GNN model and ensembled several students to construct a powerful teacher for online distillation.

Platforms and Practice of Graph Neural Networks 8

Tianyu Zhao and Yaoqi Liu

8.1   Introduction

The rapid development of deep learning in recent years benefits from some easy-to-use and efficient deep learning platforms (e.g., TensorFlow [1],[1] PyTorch [150][2]). However, graph is a kind of irregular and complex data. The platforms do not have a proper data structure for the storage and processing of a graph. The computation in deep learning platforms is usually done by the calculations on dense tensors, but adjacency matrices of graphs are usually sparse. And mini-batch training usually involves graph slicing that is inconvenient to be implemented on deep learning platforms. Thus, GNN models are hard to be implemented directly and quickly by traditional deep learning platforms which brings inconvenience to the study of GNNs.

The limitation of the deep learning platforms has delayed the development of GNNs. To resolve this limitation, some platforms have been developed to support GNN computation based on deep learning platforms. For example, DGL [193] can perform optimizations transparently by advocating graph as the central programming abstraction. It is framework agnostic that supports PyTorch, TensorFlow, and MXNet, but it needs different code implementations to be compatible with different backends. PyG [50] achieves high performance by leveraging dedicated CUDA kernels. It uses PyTorch as its backend and provides Tensor-centric API that "breaks" the graph into several key tensors, but it only supports a single backend.

---

[1] https://www.tensorflow.org/
[2] https://pytorch.org/

T. Zhao (✉) · Y. Liu
Beijing University of Posts and Telecommunications, Beijing, China

© The Author(s), under exclusive license to Springer Nature Switzerland AG 2023
C. Shi et al., *Advances in Graph Neural Networks*, Synthesis Lectures
on Data Mining and Knowledge Discovery,
https://doi.org/10.1007/978-3-031-16174-2_8

In this chapter, we will present GammaGL.[3] which is a novel GNN platform developed by BUPT GAMMA Lab.[4] It supports multiple deep learning platforms with the same code implementation. Users may choose their favorite backends from TensorFlow [1], PyTorch [150], PaddlePaddle,[5] and Mindspore.[6] GammaGL also uses Tensor-centric API, which makes it friendly to users familiar with PyG. And we will also introduce the basic usage of GammaGL with three typical GNN models (GIN [215], GraphSAGE [74], HAN [200]).

8.2   Foundation

In this section, we will first introduce deep learning platforms. Deep learning platforms have facilitated the development of many related algorithms, but do not support GNNs well. Therefore, the platforms of GNNs were created, which are built upon the deep learning platforms. We will then introduce the platforms of GNNs.

8.2.1   Deep Learning Platforms

Deep learning platforms provide users with the tools to develop and deploy deep learning algorithms. Deep learning platforms can support data management and processing, automatic differentiation, and model training and deployment. They usually support GPU devices to accelerate and optimize computation like matrix multiplication. In this part, we will introduce some widely used platforms i.e., TensorFlow [1], Pytorch [150], PaddlePaddle, MindSpore, and a framework TensorlayerX [106][7] that is compatible with multiple deep learning frameworks.

8.2.1.1 TensorFlow

TensorFlow [1] is an end-to-end open-source machine learning platform. It is initially developed by the Google Brain team for internal use in machine learning and deep neural network research. The initial version was released in 2015, and since then millions of algorithms aiming to conduct artificial intelligence research based on TensorFlow have emerged. Although the initial generation of TensorFlow was criticized as not flexible enough since it adopted static computational graph as the automatic differentiation scheme, it was still preferred in the commercial application for its convenient visualization (TensorBoard), great portability, and high efficiency. TensorFlow 2.0 version was released in 2019 and it

---

[3] https://github.com/BUPT-GAMMA/GammaGL.

[4] https://github.com/BUPT-GAMMA.

[5] https://www.paddlepaddle.org.cn/.

[6] https://www.mindspore.cn/.

[7] https://github.com/tensorlayer/TensorLayerX.

takes dynamic computational graphs as the default scheme from then on. Now, TensorFlow provides stable Python and C++ API, as well as non-guaranteed backward compatible API for JavaScript, etc. Besides, it has a comprehensive, flexible ecosystem of tools, libraries, and community resources that lets researchers push the state of the art in ML and developers easily build and deploy ML-powered applications. Therefore, it is now general enough to be applicable in a wide variety of other domains.

Beginners may be confused by the concepts of the execution modes in TensorFlow. The execution mode determines how automatic differentiation and other basic operations are performed. TensorFlow provides two alternative modes, the *eager* mode and the *graph* mode. In *eager* mode, users can run programs eagerly, which means those TensorFlow operations are executed by Python, operation by operation, and returning results to Python. This mode makes operations similar to a normal python program and is more flexible to use and easier to debug. In *graph* mode, all operations in TensorFlow will be first converted into a computational graph and tensor computations are then executed by this graph. This mode enables portability outside Python and tends to offer better performance and efficiency. By the way, $tf.function$ is provided to allow users to switch from *eager* execution to *graph* execution (Fig. 8.1).

TensorFlow has the following features:

• **Alternative Training Strategies**: TensorFlow offers multiple levels of abstraction so users can choose the right level to balance the development cost and the model performance. For beginners of TensorFlow and machine learning, high-level API (like Keras and Premade Estimators) is provided to help build and train models simply and quickly. While, for the large learning tasks in practical application, $DistributionStrategy$ API for distributed

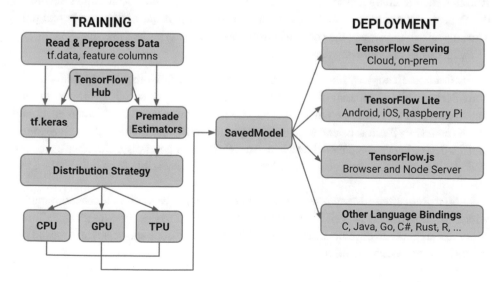

Fig. 8.1 Scheme diagram for API in TensorFlow 2.0

training is provided to automatically adapt to different hardware configurations (from CPUs, GPUs to TPUs) without changing the model definition.

• **Convenient Deployment**: TensorFlow also provides a direct and robust path to production. Owing to its good portability, whether on servers, edge devices, or the web, training and deploying models using TensorFlow can be easy, no matter what language or platform you use. If you require a full production ML pipeline, TensorFlow Extended (TFX) is needed. For running inference on mobile and edge devices, TensorFlow Lite is provided. For training and deploying models in browser and node server, TensorFlow.js is recommended. Moreover, other language bindings are also supported, including C, Java, Go, C#, Rust, R, etc.

• **Powerful Research Support**: TensorFlow makes it easy to take new ideas from concept to code, and from model to publication. Keras Functional API and Model Subclassing API allow for the creation of complex topologies including using residual layers, custom multi-input/-output models, and imperatively written forward passes. And for even more flexibility and control, the low-level TensorFlow API is always available and working in conjunction with the higher-level abstractions for fully customizable logic. TensorFlow also provides a powerful visualization toolkit and debugger, TensorBoard, for researchers.

8.2.1.2 PyTorch

PyTorch [150] is an open-source machine learning platform based on the Torch library, primarily developed by Facebook's AI Research lab (FAIR). Researchers and developers can make fast, flexible experimentation and efficient production with PyTorch, as it provides a user-friendly front-end, distributed training, and an ecosystem of tools and libraries.

PyTorch designs tensors computing (like NumPy) with strong acceleration via GPUs. It uses dynamic computation graphs and is completely Pythonic, so it allows scientists, developers, and neural network debuggers to run and test portions of the code in real time. Thus, users can implement the code and check its correctness at the same time.

PyTorch has the following features:

• **Production Ready**: PyTorch provides ease of use and flexibility in eager mode with TorchScript while seamlessly transitioning to graph mode for speed, optimization, and functionality in C++ runtime environments. The transition seamlessly between eager and graph modes. PyTorch accelerates the path to production with TorchServe, an easy-to-use tool for deploying PyTorch models at scale. And, PyTorch supports asynchronous execution of collective operations and peer-to-peer communication accessible from both Python and C++,

• **Robust Ecosystem**: Many researchers and developers have built an active community, which has contributed to a rich ecosystem of tools and libraries for extending PyTorch. More researchers and developers are using PyTorch to carry out large-scale training and run models in a production-scale environment.

- **Dynamic Neural Networks**: PyTorch is the first deep learning platform to use dynamic computational graphs. It has a unique way of building neural networks: using and replaying a tape recorder. PyTorch uses a technique called reverse-mode auto-differentiation, which allows us to change the way our network behaves arbitrarily with zero lag or overhead.

8.2.1.3 PaddlePaddle

PaddlePaddle is a deep learning platform developed by Baidu for research and applications. It is the first independent R&D (research and development) deep learning platform in China, which integrates deep learning core training and inference platform, basic model libraries, end-to-end development kits, and rich tool components.

PaddlePaddle has the following features:

- **Easy to develop**: The PaddlePaddle has an easy-to-learn and easy-to-use interface and an efficient internal core architecture. PaddlePaddle is compatible with both imperative and declarative programming paradigms. It also realizes the industry's first dynamic and static unified programming paradigm.
- **Ultra-large-scale deep learning model**: PaddlePaddle has broken through the training technology of ultra-large-scale deep learning models. It has taken the lead in realizing the ability of parallel training of 100 billion features and trillions of parameters with data sources distributed over hundreds of nodes. PaddlePaddle also overcomes the online learning and deployment challenges of ultra-large-scale deep learning models.
- **Supports both dynamic and static graph**: At the same time, PaddlePaddle provides users with two kinds of calculation graphs, dynamic graphs, and static graphs. Dynamic graph construction network is more flexible and network debugging is convenient, while static graph deployment is easy and application implementation is more efficient.

8.2.1.4 MindSpore

MindSpore is a next-generation deep learning platform that incorporates the best practices of the industry. It manifests the computing power of the Ascend AI Processor and supports flexible all-scenario deployment across device-edge-cloud. MindSpore creates a brand-new AI programming paradigm and lowers the threshold for AI development. MindSpore was launched by Huawei in August 2019 and officially opened its source on March 28, 2020. It aims to achieve easy development, efficient execution, and all-scenario coverage.

MindSpore has the following features:

- **Automatic Differentiation**: To facilitate easy development, MindSpore adopts an automatic differentiation (AD) mechanism based on source code transformation (SCT), which can represent complex combinations through control flows. A function is converted into an intermediate representation (IR) which constructs a computational graph that can be parsed and executed on devices. Before execution, multiple software and hardware collaborative optimization technologies are used in the graph to improve performance and efficiency in various scenarios across the device, edge, and cloud.

• **Dynamic Graph**: MindSpore supports dynamic graphs without introducing additional AD mechanisms (such as the operator overloading AD mechanism). This results in significantly greater compatibility between dynamic and static graphs.

• **Automatic Parallelism**: To effectively train large models on large datasets, MindSpore supports data parallel, model parallel, and hybrid parallel training through advanced manual configuration policies. In addition, MindSpore supports the automatic parallelism which efficiently searches for a fast parallel strategy in a large strategy space.

• **Device-Cloud Collaborative Architecture**: MindSpore aims to build an AI platform that covers all scenarios from the device-side to the cloud-side. MindSpore supports "device-cloud" collaboration capabilities, which include model optimization, on-device inference, and device-cloud collaborative learning.

8.2.1.5 TensorLayerX

TensorLayerX [106] is a deep learning library designed for researchers and engineers, that is compatible with multiple deep learning platforms such as TensorFlow, MindSpore, PaddlePaddle, and PyTorch(partial), allowing users to run the code on different hardware like Nvidia-GPU and Huawei-Ascend, and supports hybrid-framework programming. It provides popular *Deep Learning* and *Representation Learning* modules that can be easily customized and assembled for tackling real-world machine learning problems. This project is maintained by researchers from Peking University, Imperial College London, Princeton, Stanford, Tsinghua, Edinburgh, and Peng Cheng Lab.

TensorLayerX has the following features:

• **Efficient Compatible**: TensorLayerX is developed using pure Python code. By encapsulating multiple backend Python interfaces, TensorLayerX provides a unified API for deep learning development that is compatible with multiple platforms. Then, the underlying programs of each backend framework are responsible for calling hardware computing, enabling developers to conduct deep learning development regardless of backend frameworks and hardware platforms. In this process, there is almost no calculated performance loss.

• **Easy To Use**: First, TensorLayerX uses mainstream AI framework and chips, which can effectively reduce learning costs. Then, TensorLayerX's development paradigm is object-oriented, with all layers and models defined by inheritance and rewriting the *nn.Module* type. The bottom layer is *tlx.ops* which encapsulate the basic tensor operation of various backend framework. On this basis, by rewriting the *nn.Module* type, TensorLayerX encapsulates many commonly used neural network layers and modules, and developers can easily write their own algorithms. Finally, TensorLayerX has considered concise training process and customized training process in design. The user can either use the encapsulated *model.train*() method to start model training, or use a circular way to accurately control the training process of each step.

• **Robust Ecosystem**: TensorLayerX is not just a framework, but a deep learning platform composed of open-source products, open-source communities, and open-source activities, which form the open-source ecosystem of TensorLayerX in many ways. TensorLayerX has many derivative products, including the TLXZOO algorithm library, which covers various common neural network algorithms in computer vision, natural language processing, and other fields, making it easy for developers to reuse. There is also the RLZOO Reinforcement learning toolkit, which is a collection of the most useful reinforcement learning algorithms, frameworks, and applications.

8.2.2   Platforms of Graph Neural Networks

As we can see, the deep learning platforms are already excellent. However, applying GNNs on deep learning platforms still faces a lot of challenges. Graph data usually contains adjacency matrices with associated features, but the platforms do not have a proper data structure for the storage and processing. The adjacency matrices of graphs are usually sparse, but the computation in deep learning platforms is usually done by the calculations on dense tensors. In GNNs, due to the data dependency problem, mini-batch training usually involves graph slicing, which is also difficult to be implemented with deep learning platforms. Therefore, the platforms of GNNs are built in recent years to provide a succinct programming interface. With the platforms developing fast, they have more complete functions and are used widely. In this part, we will introduce DGL [193] and PyG [50], two mature platforms of GNNs.

8.2.2.1   DGL

DGL [193], Deep Graph Library, is a Python package specifically designed for deep learning on graphs, which is easy to use, high performance, and scalable. It is developed and maintained by Amazon Web Services AI Shanghai Lablet, Amazon Web Services Machine Learning, NVIDIA, and New York University, which is widely used by developers and researchers on GNNs.

DGL is framework agnostic, meaning if a model is one component of the end-to-end application, then the rest of the logic can be implemented in any major framework, such as PyTorch, Apache MXNet, or TensorFlow.

As shown in Fig. 8.2, DGL is graph-centric and develops various neural network modules. With user-friendly message passing interfaces, researchers can implement a graph model easily and efficiently. DGL now supports multi-GPU sampling and training, which makes more developers and researchers use it.

DGL has the following features:

• **Framework agnostic**: DGL supports multiple backends. Users may choose their favorite backend from PyTorch, TensorFlow, and MXNet and can get used to DGL within a short period of time.

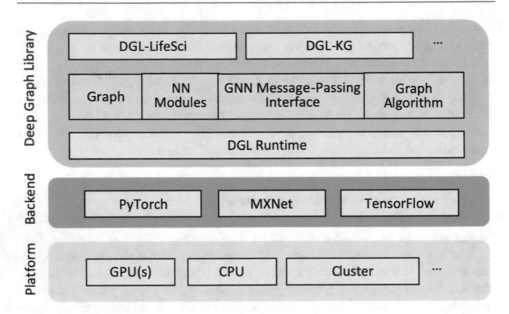

Fig. 8.2 Overview DGL Framework

• **Support GPU**: DGL provides a powerful graph object that can reside on either CPU or GPU. It combines graph data and features to control them easily. What is more, DGL provides various functions to compute graph data with GNNs efficiently.

• **A versatile tool for GNN researchers and practitioners**: DGL provides DGL-Go, which is a command-line interface to get started with training, using, and studying famous GNN models. DGL collects many models from a variety of domains, so researchers can search these models and use them as a baseline for experiments. Moreover, DGL has many state-of-the-art GNN layers and modules to build new models.

• **Scalable and efficient**: DGL now supports multiple GPUs and multiple machines for sampling or training on large-scale graphs. It can optimize the whole stack to reduce the consumption of communication, memory, and synchronization. As a result, DGL now can scale to billion-sized graphs.

Many teams have used DGL to develop excellent projects in various fields like:

• **DGL-LifeSci** [114]: a DGL-based package for various applications in life science with GNNs.

• **DGL-KE** [250]: a DGL-based package for learning large-scale knowledge graph embeddings.

• **OpenHGNN**[8]: a model zoo and benchmarks based on DGL for HGNNs (Heterogeneous Graph Neural Networks).

---

[8] https://github.com/BUPT-GAMMA/OpenHGNN.

8.2.2.2 PyG

PyG [50], PyTorch Geometric, is a library built upon PyTorch to easily write and train Graph Neural Networks (GNNs) for a wide range of applications related to structured data. PyTorch is more and more dominant in academic research due to its dynamic graph programming, PyG follows PyTorch in having this advantage as well.

PyG is friendly to those beginners who are familiar with PyTorch since its Tensor-centric API and design principles are close to vanilla PyTorch. Besides, PyG provides various methods for deep learning on graphs and other irregular structures, also known as geometric deep learning. For the large learning tasks on the graph, it provides easy-to-use mini-batch loaders on many small and single giant graphs as well as a mature distribution strategy API for distributed training on multi GPUs. PyG also provides a large number of common benchmark datasets and the GraphGym experiment manager to help researchers reproduce GNN experiments.

As shown in Fig. 8.3, PyG provides a multi-layer framework that enables users to select the proper one. *Engine* layer adopts PyTorch as its underlying deep learning framework and some efficient CUDA libraries to process sparse data. *Storage* layer handles data processing, transformation, and loading pipelines. It provides mini-batching and various graph sampling APIs to deal with large-scale data. *Operator* layer mainly focuses on the most essential part in GNNs, message passing algorithms, and some other basic graph algorithms, including clustering, pooling, and normalization. *Model* layer provides plenty of examples that showcase GNN models on standard graph benchmarks.

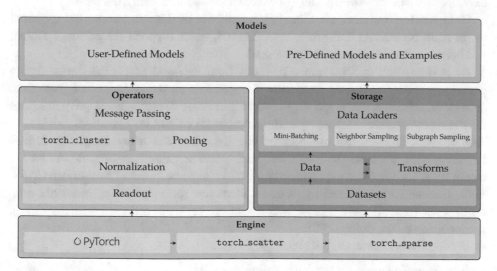

Fig. 8.3 Overview of Multi-layer PyG Framework

PyTorch Geometric has the following features:

• **Tensor-centric API**: Tensor-centric means that instead of treating the graph data as a whole, PyG "breaks" the graph into several key tensors, *edge_index*, *node_feat*, and *edge_feat*. PyG utilizes a Tensor-centric API and keeps design principles close to vanilla PyTorch, so it is friendly to those who are already familiar with PyTorch. Tensor-centric programming mode makes it easier and free to build GNN models but more difficult to process graphs with complex structures.

• **Real-world datasets and well-maintained GNN models support**: PyG provides plenty of real-world datasets from citation network to bioinformatic network. Besides, most of the state-of-the-art GNNs have been implemented by library developers or authors of research papers and are ready to be applied.

• **GraphGym integration**: GraphGym [228] enables users to easily reproduce GNN experiments, it can launch and analyze thousands of different GNN configurations. Users can customize modules by registering them to the GNN learning pipeline.

8.2.3   GammaGL

Now, we have already introduced two mature platforms of GNNs. The development of GNNs still lacks a library of algorithms based on a unified deep learning framework, which can reduce the costs associated with different learning frameworks. GammaGL, which is developed by BUPT GAMMA Lab, is developed to cope with such a dilemma.

GammaGL, Gamma Graph Library, is a multi-backend graph learning library based on the multi-backend AI framework, TensorLayerX, which supports TensorFlow, PyTorch, PaddlePaddle, and MindSpore as the backend.

As is shown in Fig. 8.4, GammaGL learns from PyG and adopts the similar design philosophy. The *Graph data* component is responsible for the abstraction of graph data, datasets management, and graph transformation, and provides storage and query functions for the upper interface. *Message passing* component utilizes the operations provided in TensorLayerX [106] to realize the underlying algorithm. *Sampling* component defines frequently used graph sampling API, like mini-batch, neighbor sampling, etc. As for the model level, GammaGL also provides the example implementation of mainstream GNNs. Thus, GammaGL is proposed by us as an alternative to implementing geometric deep learning for those who are familiar with any of the backends and prefers Tenser-centric programming mode, like PyG [50].

In conclusion, GammaGL has two major features:

• **Multi-backend Support**: GammaGL supports multiple deep learning backends, including TensorFlow, PyTorch, PaddlePaddle and MindSpore. Different from DGL framework-agnostic, the multi-backend support in GammaGL is much more powerful and convenient than that in DGL.

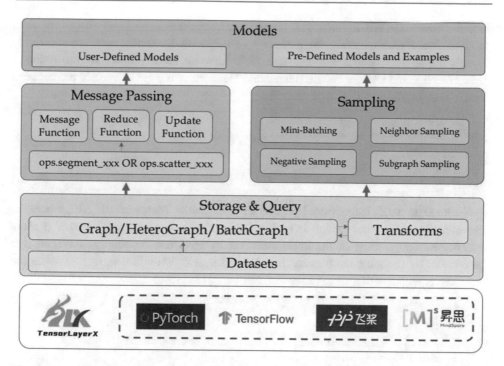

Fig. 8.4 Overview of GammaGL Framework

• **PyG-like**: Following the same design philosophy with PyG, GammaGL utilizes the Tensor-centric API. It is more flexible to customize your own message passing mechanism with the Tensor-centric API. Besides, if you are familiar with PyG, it will be friendly and maybe a TensorFlow Geometric, Paddle Geometric, or MindSpore Geometric to you.

Multi-backend Support Compared with the framework agnostic DGL [193], GammaGL is obviously much more powerful in terms of multi-backend support. DGL utilizes DLPack to share the structures of tensors among different deep learning frameworks, which makes the whole end-to-end GNN models framework agnostic (basically including the query and calculation on the DGL graph). A complete application of GNN contains loading datasets, building layers and models, and constructing trainers, most of which are frameworks specified in DGL. For example, when switching from PyTorch to TensorFlow, the codes need to be modified involving three parts:

1. Inheritance of model class (like replacing $torch.nn.Module$ with $tensorflow.keras.layer.Layer$);
2. Initialization of submodule and parameters;
3. Specific operation (like replacing $torch.matmul$ with $tensorflow.matmul$).

To further simplify, GammaGL adopts TensorLayerX to achieve the multi-backend support, where the code modification mentioned above has been realized. Therefore, the same code implementation can take different deep learning frameworks as its backend without any extra modification. Besides, considering preferences for specific backends, users are allowed to directly utilize the special operation in specific framework, namely, the mixed utilization of TensorLayerX and another certain framework. Here are two ways to specify the backend for the code implementation in GammaGL.

```
# Specify backend in command line
TL_BACKEND=paddle python gcn_trainer.py

# Or specify backend in python program
import os
os.environ['TL_BACKEND'] = 'paddle' # tensorflow, torch, mindspore
```

PyG-like GammaGL learns from PyG and also adopts the Tensor-centric API, where the representation of a whole graph is broken into different tensors, which makes it more flexible. Basically, most of the implementation by PyG can be simply converted into GammaGL by changing the backend into TensorLayerX. We will illustrate this with the code implementation of a Graph Attention Network [189] layer.

```
# GammaGL implementation
import os
# set your backend here, default 'tensorflow'
os.environ['TL_BACKEND'] = 'paddle'
import tensorlayerx as tlx
from tensorlayerx.nn import ELU, Linear, LeakyReLU
from gammagl.layers.conv import MessagePassing
from gammagl.utils import segment_softmax

class GATLayer(MessagePassing):
    def __init__(self, in_dim, out_dim):
        super().__init__()
        self.fc = Linear(out_features=out_dim, in_features=in_dim)
        self.attn_fc_src = Linear(out_features=1, in_features=in_dim)
        self.attn_fc_dst = Linear(out_features=1, in_features=in_dim)
        self.leaky_relu = LeakyReLU()

    def message(self, h, edge_index):
        node_src = edge_index[0, :]
        node_dst = edge_index[1, :]
        weight_src = tlx.gather(self.attn_fc_src(h), node_src)
        weight_dst = tlx.gather(self.attn_fc_dst(h), node_dst)
        weight = self.leaky_relu(weight_src + weight_dst)
        alpha = segment_softmax(weight, node_dst)
        h = tlx.gather(h, node_src) * tlx.expand_dims(alpha, -1)
        return h

    def forward(self, h, edge_index):
        h = self.fc(h)
        h = self.propagate(edge_index=edge_index, x=h)
        return h
```

```
# PyTorch implementation
import torch as th
from torch.nn import ELU, Linear, LeakyReLU
from torch_geometric.nn.conv import MessagePassing
from torch_geometric.utils import softmax as segment_softmax

class GATLayer(MessagePassing):
    def __init__(self, in_dim, out_dim):
        super().__init__()
        self.fc = Linear(out_features=out_dim, in_features=in_dim)
        self.attn_fc_src = Linear(out_features=1, in_features=in_dim)
        self.attn_fc_dst = Linear(out_features=1, in_features=in_dim)
        self.leaky_relu = LeakyReLU()

    def message(self, h, edge_index):
        node_src = edge_index[0, :]
        node_dst = edge_index[1, :]
        weight_src = th.gather(self.attn_fc_src(h), node_src)
        weight_dst = th.gather(self.attn_fc_dst(h), node_dst)
        weight = self.leaky_relu(weight_src + weight_dst)
        alpha = segment_softmax(weight, node_dst)
        h = th.gather(h, node_src) * th.unsqueeze(alpha, -1)
        return h

    def forward(self, h, edge_index):
        h = self.fc(h)
        h = self.propagate(edge_index=edge_index, x=h)
        return h
```

8.3   Practice of Graph Neural Networks on GammaGL

There are many applications on the graph, such as node-level tasks, edge-level tasks, and graph-level tasks. Applying GNN on the graph is the most popular way, so we will introduce how to apply a GNN with GammaGL. In this section, we will show how to build a GNN model based on GammaGL in practice. The rest of this section is organized in the following order:

• **Create your own Graph**: This part will introduce how to create a *Graph* object in GammaGL. And we will introduce how to build *gammagl.data.InMemoryDataset* object.

• **Create message-passing network**: This part will show how to implement message passing layer and provide an instance for it.

• **Advance Mini-Batching**: This part will tell you how to store multiple graphs into a single *Graph* object for graph-level tasks.

• **Examples in practice**: The last part is the implementation of three models (i.e., GIN [215], GraphSAGE [74], HAN [200]).

8.3.1   Create Your Own Graph

In order to better introduce the practice of GammaGL, we first introduce the data structure and data processing in GammaGL. A graph is used to model pairwise relations (edges) between objects (nodes). A graph in GammaGL is described by an instance of $gammagl.data.Graph$, which has the following attributes:

$Graph.x$: Node feature matrix with shape [$num\_nodes, num\_node\_features$].

$Graph.edge\_index$: Graph connectivity in COO format with shape [$2, num\_edges$] and type $int64$.

$Graph.edge\_attr$: Edge feature matrix with shape [$num\_edges, num\_edge\_features$].

Now, we will give an example of a graph created by GammaGL with three nodes and four edges. Each node contains exactly one dimension feature (Fig. 8.5).

```
import tensorlayerx as tlx
from gammagl.data import Graph

edge_index = tlx.convert_to_tensor([[0, 1, 1, 2],
                                    [1, 0, 2, 1]], dtype=tlx.int64)
x = tlx.convert_to_tensor([[-1], [0], [1]], dtype=tlx.float32)

graph = Graph(x=x, edge_index=edge_index)
>>> Graph(edge_index=[2, 4], x=[3, 1])
```

We have processed a number of graph datasets in advance and integrated into the GammaGL. If you want to customize a dataset, you should first overwrite $gammagl.data.$ $InMemoryDataset$. The main processing procedures happen in the body of $process()$. We read, create $Graph$ object lists and collate them into one huge $Graph$ object via $gammagl.data.InMemoryDataset.collate()$ before saving.

Here, we can see a simplified example:

Fig. 8.5 An example graph created by GammaGL

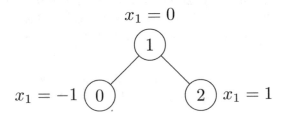

```
import tensorlayerx as tlx
from gammagl.data import InMemoryDataset, download_url

class MyOwnDataset(InMemoryDataset):
    def __init__(self, root, transform=None, pre_transform=None,
                 pre_filter=None):
        super().__init__(root, transform, pre_transform, pre_filter)
        self.data, self.slices = self.load_data(self.processed_paths[0])

    @property
    def raw_file_names(self):
        return ['some_file_1', 'some_file_2', ...]

    @property
    def processed_file_names(self):
        return tlx.BACKEND + '_data.pt'

    def download(self):
        # Download to 'self.raw_dir'.
        download_url(url, self.raw_dir)
        ...

    def process(self):
        # Read data into huge 'Data' list.
        data_list = [...]
        if self.pre_filter is not None:
            data_list = [data for data in data_list if
                         self.pre_filter(data)]
        if self.pre_transform is not None:
            data_list = [self.pre_transform(data)
                         for data in data_list]

        data, slices = self.collate(data_list)
        self.save_data((data, slices), self.processed_paths[0])
```

8.3.2   Create Message-Passing Network

This part will introduce how to create message-passing network. Generalizing the convolution operator to irregular domains is typically expressed as a neighborhood aggregation or message-passing scheme.

GammaGL provides the *Message Passing* base class, which helps in creating such kinds of message passing GNNs by automatically taking care of message propagation.

For example, a *GCNConv* can be created in the following steps:

1. Add self-loops to the adjacency matrix.
2. Linearly transform node feature matrix.
3. Compute normalization coefficients.
4. Normalize node features.
5. Sum up neighboring node features ("*sum*" aggregation).

Steps 1–3 are typically computed before message passing takes place. Steps 4–5 can be easily processed by overwriting the *Message Passing* base class. The full layer implementation is shown below:

```python
import tensorlayerx as tlx
from gammagl.layers.conv import MessagePassing
from gammagl.utils import add_self_loops, degree
from gammagl.mpops import unsorted_segment_sum

class GCNConv(MessagePassing):
    def __init__(self, in_channels, out_channels, add_bias):
        super().__init__()
        self.lin = tlx.layers.Linear(in_channels, out_channels)

    def forward(self, x, edge_index):
        # x has shape [N, in_channels]
        # edge_index has shape [2, E]

        # Step 1: Add self-loops to the adjacency matrix.
        edge_index, _ = add_self_loops(edge_index, num_nodes=x.shape[0])

        # Step 2: Linearly transform node feature matrix.
        x = self.lin(x)

        # Step 3: Compute edge weight.
        src, dst = edge_index[0], edge_index[1]
        edge_weight = tlx.ones(shape=(edge_index.shape[1], 1))
        deg = degree(dst, num_nodes=x.shape[0])
        deg_inv_sqrt = tlx.pow(deg, -0.5)
        weights = tlx.ops.gather(deg_inv_sqrt, src) * \
        tlx.reshape(edge_weight,
                    (-1,)) * tlx.ops.gather(deg_inv_sqrt, dst)

        # Step 4-5: Start propagating messages.
        return self.propagate(x, edge_index, edge_weight=weights,
                              num_nodes=x.shape[0], aggr_type='sum')

    def message(self, x, edge_index, edge_weight):
        msg = tlx.gather(x, edge_index[0, :])
        return msg * edge_weight
```

8.3.3 Advanced Mini-Batching

This part will introduce advanced mini-batching in GammaGL. The creation of mini-batching is crucial for letting the training of a deep learning model scale to huge amounts of data. Instead of processing examples one by one, a mini-batch groups a set of examples into a unified representation where it can efficiently be processed in parallel.

GammaGL opts for a new approach to achieving parallelization across a number of examples. Here, adjacency matrices are stacked in a diagonal fashion (creating a giant graph that holds multiple isolated subgraphs), and node and target features are simply concatenated in the node dimension, i.e.,

$$A = \begin{bmatrix} \mathbf{A}_1 & & \\ & \ddots & \\ & & \mathbf{A}_n \end{bmatrix}, \quad X = \begin{bmatrix} \mathbf{X}_1 \\ \vdots \\ \mathbf{X}_n \end{bmatrix}, \quad Y = \begin{bmatrix} \mathbf{Y}_1 \\ \vdots \\ \mathbf{Y}_n \end{bmatrix}.$$

This procedure has some crucial advantages over other batching procedures:

1. GNN operators that rely on a message-passing scheme do not need to be modified since messages are still hard to be exchanged between two nodes that belong to different graphs.
2. There is no computational or memory overhead. For example, this batching procedure works completely without any padding of node or edge features. Note that there is no additional memory overhead for adjacency matrices since they are saved in a sparse fashion holding only non-zero entries, i.e., the edges.

GammaGL automatically takes care of batching multiple graphs into a single giant graph with the help of the *gammagl.loader.DataLoader* class. Internally, *DataLoader* is just a regular *TensorLayerx.tensorlayerx.dataflow.DataLoader* that overwrites its *collate*() functionality, i.e., the definition of how a list of examples should be grouped. Therefore, all arguments that can be passed to a TensorLayerX *DataLoader* can also be passed to a GammaGL *DataLoader*, e.g., the number of workers *num_workers*.

In its most general form, the GammaGL *DataLoader* will automatically increment the *edge_index* tensor by the cumulated number of nodes of all graphs that got collated before the currently processed graph and will concatenate *edge_index* tensors (that are of shape [2, *num_edges*]) in the second dimension. The same is true for *face* tensors, i.e., face indices in meshes. All other tensors will just get concatenated in the first dimension without any further increment in their values.

However, there are a few special use cases where the user actively wants to modify this behavior to its own needs. GammaGL allows modification to the underlying batching procedure by overwriting the *gammagl.data.Graph.__inc__*() and *gammagl.data.Graph.__cat_dim__*() functionalities.

8.3.4   Practice of GIN

This part will introduce the practice of GIN. Graph Isomorphism Network (GIN [215]) generalizes the WL test, whose discriminative/representational power is equal to the power of the WL test, and hence achieves maximum discriminative power among GNNs.

GIN updates node representations as

$$h_i^{(l+1)} = f_{\Theta}\left((1+\epsilon)h_i^l + \text{aggregate}\left(\left\{h_j^l, j \in \mathcal{N}(i)\right\}\right)\right). \tag{8.1}$$

To consider all structural information, we use information from all depths/iterations of the model. GIN achieves this by using graph representations concatenated across all iterations/layers of GIN:

$$h_G = CONCAT \left(READOUT \left(\{h_v^{(k)} | v \in G\} \right) \middle| k = 0, 1, ..., K \right) \qquad (8.2)$$

GIN deals with graph-level tasks and the input dataset contains multiple graphs. It requires batching multiple graphs into a single giant graph using *gammagl.loader.DataLoader* class.

```
from gammagl.loader import DataLoader

dataset = TUDataset(args.dataset_path, args.dataset, use_node_attr=True)
train_dataset = dataset[:110]
train_loader = DataLoader(dataset=train_dataset, batch_size=64,
                          shuffle=False)
```

Next comes the code for *GINConv*, which implements formula (8.1).

```
class GINConv(MessagePassing):
    def __init__(self,
                 apply_func=None,
                 aggregator_type='mean',
                 init_eps=0.0,
                 learn_eps=False):
        super(GINConv, self).__init__()
        self.apply_func = apply_func
        if aggregator_type in ['mean', 'sum', 'max']:
            self.aggregator_type = aggregator_type
        else:
            raise KeyError('Aggregator type {} not recognized.'
                           .format(aggregator_type))

        # to specify whether eps is trainable or not.
        # self.eps = tlx.Variable(initial_value=[float(init_eps)],
        #                         name='eps', trainable=learn_eps)
        init = tlx.initializers.Constant(value=init_eps)
        self.eps = self._get_weights(var_name='eps',
                                     shape=(1,), init=init)

    def forward(self, x, edge_index):
        out = self.propagate(x=x, edge_index=edge_index,
                             aggr=self.aggregator_type)

        out += (1 + self.eps) * x

        if self.apply_func is not None:
            out = self.apply_func(out)
        return out
```

The *apply_func* in *GINConv* is a callable activation function or activation layer. If *apply_func* is not *None*, apply this function to the updated node feature, equivalent to f_Θ in the formula (8.1).

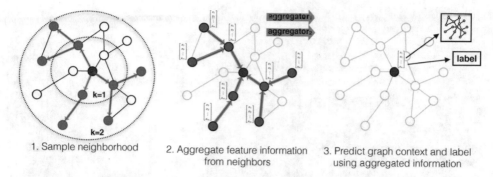

1. Sample neighborhood

2. Aggregate feature information from neighbors

3. Predict graph context and label using aggregated information

Fig. 8.6 Visual illustration of the GraphSAGE sample and aggregate approach

```python
class ApplyNodeFunc(nn.Module):
    """Update the node feature hv with MLP, BN and ReLU."""

    def __init__(self, mlp):
        super(ApplyNodeFunc, self).__init__()
        self.mlp = mlp
        self.bn = nn.BatchNorm1d(num_features=self.mlp.output_dim)
        self.act = nn.ReLU()

    def forward(self, h):
        h = self.mlp(h)
        h = self.bn(h)
        h = self.act(h)
        return h
```

8.3.5   Practice of GraphSAGE

This part will introduce the practice of GraphSAGE. The method of GraphSAGE comes from the paper Inductive Representation Learning on Large Graphs [74]. The core idea is sampling and aggregation (Fig. 8.6).

Thus, in the practice of GraphSAGE, we need to consider how to implement sampling and layer-by-layer aggregation in GammaGL. For sampling, GammaGL uses the class $Neighbor - Sampler$ to complete the sampling of the destination nodes and obtain source nodes (neighbor nodes). Finally, we will get all sampled nodes and renumbering $edge\_index$ of sampled nodes. We take the use of sampling in the GraphSAGE model as an example, here is our core code of sampling:

```
"""
The loader realize neighbor sample,
and return subgraph, all sampled nodes, destination nodes"""
import tensorlayerx as tlx
from gammagl.loader.Neighbour_sampler import Neighbor_Sampler
train_loader = Neighbor_Sampler(edge_index=graph.edge_index.numpy(),
                                dst_nodes=tlx.convert_to_numpy(train_idx),
                                sample_lists=[25, 10], batch_size=2048,
                                shuffle=True, num_workers=6)

net = GraphSAGE_Model(args)
"""train one epoch"""
for dst_node, adjs, all_node in train_loader:
    net.set_train()
    # input : sampled subgraphs, sampled node's feat
    data = {"x": tlx.gather(x, tlx.convert_to_tensor(all_node)),
            "y": y,
            "dst_node": tlx.convert_to_tensor(dst_node),
            "subgs": adjs}
    # label is not used
    train_loss = train_one_step(data, label=tlx.convert_to_tensor([0]))
```

For aggregation, take the *mean* aggregator function in supported 4 aggregator functions as an example. The *mean* function is as follows:

$$\mathbf{h}_v^k \leftarrow \sigma \left(\mathbf{W} \cdot \text{MEAN} \left(\{ \mathbf{h}_v^{k-1} \} \cup \{ \mathbf{h}_u^{k-1}, \forall u \in \mathcal{N}(v) \} \right) \right)$$

Here is our *mean* aggregator function code:

```python
class SAGEConv(MessagePassing):
    def __init__(self, in_channels, out_channels,
                 activation=None, aggr="mean",
                 add_bias=True):
        super(SAGEConv, self).__init__()
        self.act = activation
        self.in_feat = in_channels
        # relu use he_normal
        initor = tlx.initializers.he_normal()
        # self and neighbor
        self.fc_neigh = tlx.nn.Linear(in_features=in_channels,
                                      out_features=out_channels,
                                      W_init=initor, b_init=None)

        self.fc_self = tlx.nn.Linear(in_features=in_channels,
                                     out_features=out_channels,
                                     W_init=initor, b_init=None)
        self.add_bias = add_bias
        if add_bias:
            init = tlx.initializers.zeros()
            self.bias = self._get_weights("bias",
                                          shape=(1, out_channels),
                                          init=init)
    def forward(self, feat, edge):
        if isinstance(feat, tuple):
            src_feat = feat[0]
            dst_feat = feat[1]
        else:
            src_feat = feat
            dst_feat = feat
        num_nodes = int(dst_feat.shape[0])

        src_feat = self.fc_neigh(src_feat)
        out = self.propagate(src_feat, edge,
                             edge_weight=None,
                             num_nodes=num_nodes, aggr='mean')
        out += self.fc_self(dst_feat)
        if self.add_bias:
            out += self.bias
        if self.act is not None:
            out = self.act(out)
        return out
```

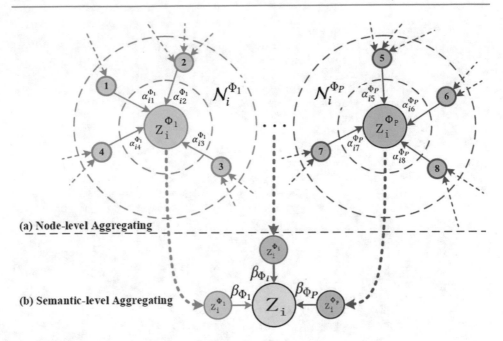

Fig. 8.7 Aggregation process of HAN

8.3.6 Practice of HAN

This part will introduce the practice of HAN. As shown in Fig. 8.7, HAN [200] firstly proposes an HGNN based on hierarchical attention, including node-level and semantic-level attentions. It enables the GNN to be directly applied to the heterogeneous graphs and further facilitates the heterogeneous graph-based applications (Fig. 8.7).

The node-level attention aims to learn the importance between a node and its meta-path-based neighbors. The mathematical formulas are as follows:

$$\alpha_{ij}^{\Phi} = \text{softmax}_j \left(e_{ij}^{\Phi} \right) = \frac{\exp \left(\sigma \left(\mathbf{a}_{\Phi}^{\mathrm{T}} \cdot \left[\mathbf{h}_i' \| \mathbf{h}_j' \right] \right) \right)}{\sum_{k \in \mathcal{N}_i^{\Phi}} \exp \left(\sigma \left(\mathbf{a}_{\Phi}^{\mathrm{T}} \cdot \left[\mathbf{h}_i' \| \mathbf{h}_k' \right] \right) \right)}, \tag{8.3}$$

$$\mathbf{z}_i^{\Phi} = \prod_{k=1}^{K} \sigma \left(\sum_{j \in \mathcal{N}_i^{\Phi}} \alpha_{ij}^{\Phi} \cdot \mathbf{h}_j' \right). \tag{8.4}$$

The semantic-level attention is able to learn the importance of different meta-paths. The mathematical formula is as follows:

$$\mathbf{Z} = \mathcal{F}_{att}(Z^{\Phi_1}, Z^{\Phi_2}, ..., Z^{\Phi_P}). \tag{8.5}$$

Following the hierarchical design of HAN, we implement $HANConv$ in a hierarchical way. Since the node-level attention part can be implemented by $GATConv$, we just need to implement the semantic-level attention. The following snippets are implementation of semantic-level attention:

```
class SemAttAggr(nn.Module):
    def __init__(self, in_size, hidden_size):
        super().__init__()
        self.project = Sequential(
            Linear(in_features=in_size, out_features=hidden_size),
            Tanh(),
            Linear(in_features=hidden_size, out_features=1, b_init=None)
        )

    def forward(self, z):
        w = tlx.reduce_mean(self.project(z), axis=1)
        beta = tlx.softmax(w, axis=0)
        beta = tlx.expand_dims(beta, axis=-1)
        return tlx.reduce_sum(beta * z, axis=0)
```

In the implementation of $HANConv$, we need to create $GATConv$ instances according to the meta-path and a $SemAttAggr$ instance. Following are the main part of the implementation:

```python
class HANConv(MessagePassing):
    def __init__(self, in_channels, out_channels, metadata,
                 heads=1, negative_slope=0.2, dropout_rate=0.5):
        super().__init__()
        if not isinstance(in_channels, dict):
            in_channels = {
                node_type: in_channels for node_type in metadata[0]
            }

        self.gat_dict = ModuleDict({})
        for edge_type in metadata[1]:
            src_type, _, dst_type = edge_type
            edge_type = '__'.join(edge_type)
            self.gat_dict[edge_type] = GATConv(
                in_channels=in_channels[src_type],
                out_channels=out_channels,
                heads=heads,
                dropout_rate=dropout_rate,
                concat=True)

        self.sem_att_aggr = SemAttAggr(in_size=out_channels*heads,
                                       hidden_size=out_channels)

    def forward(self, x_dict, edge_index_dict, num_nodes_dict):
        out_dict = {}
        # Iterate over node types:
        for node_type, x_node in x_dict.items():
            out_dict[node_type] = []

        # node level attention aggregation
        for edge_type, edge_index in edge_index_dict.items():
            src_type, _, dst_type = edge_type
            edge_type = '__'.join(edge_type)
            out = self.gat_dict[edge_type](x_dict[dst_type],
                                           edge_index,
                                           num_nodes_dict[dst_type])
            out = tlx.relu(out)
            out_dict[dst_type].append(out)

        # semantic attention aggregation
        for node_type, outs in out_dict.items():
            outs = tlx.stack(outs)
            out_dict[node_type] = self.sem_att_aggr(outs)

        return out_dict
```

8.4 Conclusion

Researchers and developers can implement GNNs in a much easier and more efficient with the platforms of GNNs. This chapter first introduces the deep learning platforms and then introduces the platforms of GNNs with some mature platforms like DGL and PyG. We also present the novel platform called GammaGL, which can support multi-backend with the same code implementation. And, we give the practice of GammaGL with three typical GNN models. The GammaGL is still under development and new features are coming soon. For more details about GammaGL, please visit the website https://github.com/BUPT-GAMMA/GammaGL.

Future Direction and Conclusion

9

9.1 Future Direction

9.1.1 Self-supervised Learning on Graphs

Recently, GNNs are argued that have heavy label reliance, since most of the works have focused on (semi-) supervised learning, in which a sufficient amount of input data and label pairs are needed. However, since a large number of labels are required, the supervised training becomes inapplicable in practical scenarios, where labels are expensive, limited, or even unavailable. To address these issues, Graph Self-supervised Learning (GSSL), which extracts informative knowledge through well-designed pretext tasks without relying on manual labels, has become a promising and trending learning paradigm for graph data.

Specifically, according to the self-supervised training schemes, the GSSL can be categorized into three classes [121, 211], including unsupervised approach, pretraining, and auxiliary training. For the data with no labels available, GSSL serves as an unsupervised approach [30]. For the data with a limited number of labeled data, GSSL serves either as a pretraining process [123] (labeled data for fine-tuning the pretrained models for downstream tasks) or as an auxiliary training task [236] (jointly trained with downstream tasks).

Recently, these methods have been applied to real world, for example, recommender systems [76] and chemistry [28]. However, the most current study merely concentrates on basic graph data, indicating that GSSL holds untapped potential for more complex graphs (heterogeneous/heterophily/directed graphs). For example, heterogeneous graph, with exclusive high-order semantic structure and GNN model design, cannot directly use the current GSSL. Moreover, it is also promising to extend GSSL to more expansive fields of applications, for instance, financial networks, cyber security, and federated learning. Lastly, existing GSSL methods are mainly designed with intuition and evaluated by empirical experiment, and thus lack the theoretical understanding [121]. It is vital to set up a solid theoretical foundation for GSSL.

9.1.2 Robustness

In recent years, deep neural network methods are often criticized for their lack of adversarial robustness [64]. Many researchers have noticed that DNNs can be easily fooled/attacked by slight deliberate perturbations of an input, also known as adversarial examples.

As the extension of deep learning on graph, graph neural networks are also proved to be fragile to adversarial attacks. Compared with previous adversarial analysis in non-graph data, the study on graph data raises unique challenges. First, it is hard to design efficient algorithms to generate adversarial examples in such discrete spaces. Second, adversarial perturbations are designed to be imperceptible to humans in the image domain, so one can force a certain distance function, such as L_p norm distance to be small between adversarial and benign instances. However in graph data, how to define imperceptible or subtle perturbation requires further analysis, measurement, and study.

Recently, some studies are proposed and proved that GNNs often suffer from adversarial attacks. Specifically, [35, 236] point out the vulnerability of GNNs to topology/feature/label attacks. This is highly critical, since especially in domains where GNNs used (e.g., the web) adversaries are common and false data is easy to inject: spammers add wrong information to social networks; fraudsters frequently manipulate online reviews and product websites [259]. Such vulnerabilities under adversarial attacks are major obstacles for GNNs to be used in the safety-critical scenarios, and thus have received increasing attention from both academia and industry. However, the majority of current adversarial robustness GNNs assume homogeneous graphs. It is difficult to directly apply current attack/defense to heterogeneous GNNs, since their inputs may contain different types of nodes and edges, or different forms of node and edge inputs, such as images and text. Therefore, new methods should be developed to handle heterogeneous graphs. Moreover, how to enhance the robustness of heterogeneous GNNs needs to be further studied.

9.1.3 Explainability

Deep neural networks, as the black-box model, a major limitation is that they are not amenable to explainability [13], namely, they are difficult to diagnose what aspects of the model's input drive the decisions. In order to safely and trustfully deploy deep models, it is necessary to provide both accurate predictions and human-intelligible explanations, especially for users in interdisciplinary domains.

In recent years, GNNs have become increasingly popular since many real-world data sets are represented as graphs, especially for interdisciplinary biochemical areas (e.g., chemical molecule and protein-protein interaction graph), where many unsolved mysteries and the domain knowledge are still lacking. These facts raise the need of developing explanation techniques to explain GNNs. While the explainability of GNNs has more challenges. Unlike images and texts, graphs contain the coupled topology and feature information as input to be explained. First, the discreteness property of topology makes the optimization of explanations intractable. Furthermore, understanding the semantic meaning of input image/text is simple and straightforward, but topology information is less intuitive for humans, since graphs can represent complex data, such as molecules, social networks, and citation networks.

Recently, several approaches [224, 229] are proposed to explain the predictions of GNNs from different angles and provide instance/model level explanations. Specifically, instance-level methods provide input-dependent explanations for each input graph, and model-level methods explain graph neural networks without respect to any specific input example. However, learning the explanation process can be difficult as no ground-truth explanations exist, and different application scenarios require diverse types of explanations. Moreover, the current evaluation of GNN explainability methods mainly relies on human cognition and quantitative analysis. So, it is necessary to build a unified treatment, a standard benchmark, and the testbed for evaluations.

9.1.4   Fairness

Deep learning is increasingly being used in high-stake decision-making applications that affect individual lives. However, deep learning models might exhibit algorithmic discrimination behaviors with respect to protected groups, potentially posing negative impacts on individuals and society [43]. Therefore, fairness in deep learning has attracted tremendous attention recently.

GNNs are the extension of deep learning on graph, but few works have been dedicated to creating unbiased GNNs. Conventional studies of fairness overwhelmingly focus on independently and identically distributed (i.i.d) data, which cannot be directly applied on graph data for the absence of simultaneous consideration of the bias from node attributes and graph structures.

Recently, several recent works [33] have proved that the fairness problem can be further amplified in GNNs, where the graph topology is shown to exhibit different biases. They refer that GNNs employ message-passing with features aggregated from neighbors, which might further intensify this bias. For example, in social networks, communities are usually more connected between themselves. They argue that such unfairness of GNNs will lead to more severe bias in decision, weaken the potential of an individual from a discriminated community. It would largely limit the wide adoption of GNNs in domains such as ranking of job applicants, crime rate prediction, and credit score estimation. Thus, it is critically important to investigate fair GNNs. As an emerging topic, it leaves many open questions. First, the investigation of intersectional fairness on graphs, i.e., combination and propagation of multiple sensitive attributes, is crucial and lacking in fair GNNs. Moreover, the removal of bias could possibly hurt the models' ability for the main prediction task of GNNs. It remains a challenge to simultaneously reduce bias and maintain satisfactory GNNs prediction performance.

9.1.5 Biochemistry

In the field of biochemistry, GNNs are applied to study the graph structure of molecules/compounds recently. Generally, the molecule or chemical compound can be modeled as graphs, where atoms are treated as nodes and chemical bonds are regarded as edges. The protein-protein interactions (PPIs) are also denoted graph, preserving the physical contacts established between multiple proteins. Moreover, in healthcare, drug-drug interactions (DDIs) also can be modeled as graphs for modeling polypharmacy side effects, where each side effect is an edge of a different type.

Based on these biochemistry graphs, GNNs show strong performance and have revolutionized many tasks of these areas, ranging from predicting molecular properties [61], inferring protein interfaces [51], and identifying the interactions of drug and target [73]. Specifically, predicting the properties of novel molecules is important for applications in material designs and drug discovery. Predicting the interface where these interactions occur is a challenging problem with important applications in drug discovery and design. The identification of drug-target interactions is vital, where researchers often use GNN applications to see which molecules are strong candidates for future drugs. These uses of GNNs could greatly improve the time consuming and costly development of material designs and drug discovery/re-purposing.

9.1.6 Physics

The data sets in particle physics are often represented by graphs, and thus have benefited from representing the interactions between particles or molecules as a graph and then predicting with GNNs. Specifically, researchers often apply GNN to simulate the dynamics of complex systems of particles by predicting at each step a relative movement of each particle. Based on it, they can reconstruct plausibly the dynamics of the whole system and further gain insights into the underlying laws governing the motion. Furthermore, researchers also utilize GNNs at the Large Hadron Collider to process millions of images and select those that could be relevant to the discovery of new particles. Besides, researchers also utilize GNNs to find new low-cost catalysts for new ways to store renewable energy, e.g., sun or wind. The catalysts can be used to drive the chemical reactions at a high rate for converting such energy into other fuels, for example, hydrogen. And the use of GNNs would reduce currently costly simulations (from days to milliseconds). More details and applications of GNNs in particle physics are described in [169].

9.2 Conclusion

In conclusion, deep learning on graphs, especially graph neural networks, is a promising and fast-developing research field that offers exciting opportunities and presents many challenges. In recent years, GNNs have significantly facilitated graph analysis and related applications. This book conducts a comprehensive study of the state-of-the-art graph neural network methods. We mainly present the frontier graph neural networks: besides the homogeneous, static, Euclidean graph neural networks methods. We also review the heterogeneous graph neural networks, the dynamic graph neural network methods, and the hyperbolic graph neural network methods.

In Chap. 1, we first briefly introduce the concepts of graphs, the matrix representations of graphs. And then we introduce a variety of complex graphs and representative computational tasks on graphs, including both node-focused and graph-focused tasks. Furthermore, we summarize the development and history of graph representation learning, especially for the graph neural networks. Finally, we briefly introduce the organization of this book.

In Chap. 2, we mainly introduce the most representative graph neural networks. We first introduce the typical graph convolutional network (GCN) from the view of spectral, then we provide the flexible and efficient spatial-based variants of graph neural networks, including GraphSAGE, GAT, and HAN.

In Chap. 3, we introduce several novel message-passing-based homogeneous graph neural networks. Specifically, we first introduce an adaptive multi-channel graph convolutional network (AM-GCN) to adaptively aggregate and fuse the feature and structure information in the message-passing process. Then we introduce a frequency adaption graph convolutional network (FAGCN) to adaptively aggregate the low-frequency and high-frequency information. Furthermore, we mention a graph estimation neural network (GEN) which can learn a better message passing structure to increase the robustness in denoising and community detection during graph convolution. Finally, we introduce a unified framework of existing GNNs, which can summarize different message-passing functions into a closed-form objective.

In Chap. 4, we introduce three representative heterogeneous graph neural networks (HGNN) dedicated to two key issues, i.e., deep degradation phenomenon and discriminative power. Specifically, HPN theoretically analyzes the deep degradation phenomenon in HGNNs and proposes a novel convolution layer to alleviate this semantic confusion; HDE injects heterogeneous distance encoding into aggregation; and HeCo employs a cross-view contrastive mechanism to capture both local and high-order structures simultaneously.

In Chap. 5, we focus on the graph neural networks that model dynamic graphs with multiple temporal interactions. We specifically introduce three dynamic graph neural networks for temporal modeling of evolving structures, including simple homogeneous topologies and temporal heterogeneous graphs. M^2DNE models both micro- and macro-dynamics of edges based on temporal point process; HPGE preserves both semantics and dynamics by learning the formation process of all heterogeneous temporal events; and DyMGNN designs dynamic meta-path and heterogeneous mutual evolution attention mechanisms to effectively capture the dynamic semantics and model the mutual evolution of different semantics.

Compared with Euclidean geometry, hyperbolic geometry can also provide a more powerful ability to embed graphs with a scale-free or hierarchical structure. As a consequence, some recent efforts begin to design GNNs in hyperbolic spaces. In Chap. 6, we introduce three hyperbolic GNNs, which learn hyperbolic graph representations to get better performance. Hyperbolic graph Attention network (HAT) is designed to learn hyperbolic graph representations in hyperbolic spaces based on attention mechanism; LGCN is proposed to guarantee the learned node features follow the hyperbolic geometries; and HHNE is proposed to preserve the structure and semantic information of HG in hyperbolic spaces.

In Chap. 7, we introduce three typical knowledge distillation frameworks for graph neural networks. We first introduce a Knowledge distillation-based framework (CPF) which combines label propagation and feature transformation to naturally preserve structure-based and feature-based prior knowledge, and improve the performance of any GNNs therefore; then we introduce LTD that are proposed to improve distillation quality for GNN models by learning node-specific temperatures. We also introduce a data-free adversarial knowledge distillation framework (named DFAD-GNN), which could distill teacher model without training data.

After making thorough discussions and summarizations of the reviewed methods, this book also discusses the future development directions for graph neural networks. Actually, there are also other ongoing or future research directions which are also worthy of pay attention. In general, studying graph neural networks or even deep learning on graphs constitutes a critical building block in modeling relational data, and it is an important step toward a future with better machine learning and artificial intelligence techniques. We especially need to pay more attention to their practical problem and broader applications, such as graph self-supervised learning, robustness, explainability, fairness, and the promising applications in biochemistry and physics areas and so on.

References

1. Abadi, M., Barham, P., Chen, J., Chen, Z., Davis, A., Dean, J., Devin, M., Ghemawat, S., Irving, G., Isard, M., et al. (2016). Tensorflow: A system for large-scale machine learning. In *OSDI* (pp. 265–283).
2. Abu-El-Haija, S., Perozzi, B., Kapoor, A., Alipourfard, N., Lerman, K., Harutyunyan, H., Steeg, G. V., & Galstyan, A. (2019). Mixhop: Higher-order graph convolutional architectures via sparsified neighborhood mixing. In *ICML* (pp. 21–29).
3. Albert, R., DasGupta, B., & Mobasheri, N. (2014). Topological implications of negative curvature for biological and social networks. *Physical Review E, 89*(3), 032811.
4. Arjovsky, M., & Bottou, L. (2017). Towards principled methods for training generative adversarial networks. arXiv:1701.04862.
5. Bachmann, G., Becigneul, G., & Ganea, O. (2020). Constant curvature graph convolutional networks. In *ICML* (pp. 486–496).
6. Balazevic, I., Allen, C., & Hospedales, T. M. (2019). Multi-relational poincaré graph embeddings. In H. M. Wallach, H. Larochelle, A. Beygelzimer, F. d'Alché-Buc, E. B. Fox, & R. Garnett (Eds.), *Advances in Neural Information Processing Systems 32: Annual Conference on Neural Information Processing Systems 2019, NeurIPS 2019, December 8–14, 2019, Vancouver, BC, Canada* (pp. 4465–4475).
7. Balcilar, M., Renton, G., Héroux, P., Gaüzère, B., Adam, S., & Honeine, P. (2020). Bridging the gap between spectral and spatial domains in graph neural networks. arXiv:2003.11702.
8. Bo, D., Wang, X., Shi, C., & Shen, H. (2021). Beyond low-frequency information in graph convolutional networks. In *AAAI* (pp. 3950–3957). AAAI Press.
9. Bojchevski, A., & Günnemann, S. (2018). Deep gaussian embedding of graphs: Unsupervised inductive learning via ranking. In *ICLR*.
10. Bonnabel, S. (2013). Stochastic gradient descent on riemannian manifolds. *IEEE Transactions on Automatic Control, 58*(9), 2217–2229.
11. Bronstein, M. M., Bruna, J., LeCun, Y., Szlam, A., & Vandergheynst, P. (2017). Geometric deep learning: going beyond euclidean data. *IEEE Signal Processing Magazine, 34*(4), 18–42.
12. Bruna, J., Zaremba, W., Szlam, A., & LeCun, Y. (2014). Spectral networks and locally connected networks on graphs. In *ICLR*.

© The Editor(s) (if applicable) and The Author(s), under exclusive license to Springer
Nature Switzerland AG 2023
C. Shi et al., *Advances in Graph Neural Networks*, Synthesis Lectures
on Data Mining and Knowledge Discovery,
https://doi.org/10.1007/978-3-031-16174-2

13. Buhrmester, V., Münch, D., & Arens, M. (2019). Analysis of explainers of black box deep neural networks for computer vision: A survey. arXiv:1911.12116.

14. Cai, B., Xiang, Y., Gao, L., Zhang, H., Li, Y., & Li, J. (2022). Temporal knowledge graph completion: A survey. arXiv:2201.08236.

15. Cai, H., Zheng, V. W., & Chang, K. (2018). A comprehensive survey of graph embedding: problems, techniques and applications. *IEEE Transactions on Knowledge and Data Engineering*.

16. Cai, L., Chen, Z., Luo, C., Gui, J., Ni, J., Li, D., & Chen, H. (2021). Structural temporal graph neural networks for anomaly detection in dynamic graphs. In *CIKM* (pp. 3747–3756).

17. Cannon, J. W., Floyd, W. J., Kenyon, R., Parry, W. R., et al. (1997). Hyperbolic geometry. In *Flavors of geometry* (Vol. 31, pp. 59–115).

18. Cao, S., Lu, W., & Xu, Q. (2015). Grarep: Learning graph representations with global structural information. In J. Bailey, A. Moffat, C. C. Aggarwal, M. de Rijke, R. Kumar, V. Murdock, T. K. Sellis, & J. X. Yu (Eds.), *Proceedings of the 24th ACM International Conference on Information and Knowledge Management, CIKM 2015, Melbourne, VIC, Australia, October 19–23, 2015* (pp. 891–900). ACM.

19. Cen, Y., Zou, X., Zhang, J., Yang, H., Zhou, J., & Tang, J. (2019). Representation learning for attributed multiplex heterogeneous network. In *KDD* (pp. 1358–1368).

20. Chamberlain, B. P., Hardwick, S. R., Wardrope, D. R., Dzogang, F., Daolio, F., & Vargas, S. (2019). Scalable hyperbolic recommender systems. arXiv:1902.08648.

21. Chami, I., Ying, Z., Ré, C., & Leskovec, J. (2019). Hyperbolic graph convolutional neural networks. *Advances in neural information processing systems 32*.

22. Chang, C.-C., & Lin, C.-J. (2011). Libsvm: A library for support vector machines. *ACM Transactions on Intelligent Systems and Technology (TIST), 2*(3), 1–27.

23. Chen, M., Wei, Z., Huang, Z., Ding, B., & Li, Y. (2020). Simple and deep graph convolutional networks. arXiv:2007.02133.

24. Chen, T., Kornblith, S., Norouzi, M., & Hinton, G. E. (2020). A simple framework for contrastive learning of visual representations. In *ICML* (pp. 1597–1607).

25. Chen, X., Yu, G., Wang, J., Domeniconi, C., Li, Z., & Zhang, X. (2019). Activehne: Active heterogeneous network embedding. In S. Kraus (Ed.), *Proceedings of the Twenty-Eighth International Joint Conference on Artificial Intelligence, IJCAI 2019, Macao, China, August 10–16, 2019* (pp. 2123–2129). ijcai.org.

26. Chen, Y., Bian, Y., Xiao, X., Rong, Y., Xu, T., & Huang, J. (2021). On self-distilling graph neural network. In *Proceedings of the Thirtieth International Joint Conference on Artificial Intelligence, IJCAI* (pp. 2278–2284).

27. Chen, Y., Wu, L., & Zaki, M. J. (2019). Deep iterative and adaptive learning for graph neural networks. arXiv:1912.07832.

28. Cheng, S., Zhang, L., Jin, B., Zhang, Q., Lu, X., You, M., & Tian, X. (2021). Graphms: Drug target prediction using graph representation learning with substructures. *Applied Sciences, 11*, 3239.

29. Cheng, Y., Yuxin, G., Chuan, S., Jiawei, L., Chunchen, W., Yao, X., Xin, L., Ning, G., & Hongzhi, Y. (2023). Learning to distill graph neural networks. In *WSDM*.

30. Chu, G., Wang, X., Shi, C., & Jiang, X. (2021). Cuco: Graph representation with curriculum contrastive learning. In *IJCAI*.

31. Chung, F. R., & Graham, F. C. (1997). *Spectral graph theory*. Providence: American Mathematical Society.

32. Cui, P., Wang, X., Pei, J., & Zhu, W. (2018). A survey on network embedding. *IEEE Transactions on Knowledge and Data Engineering*.

33. Dai, E., & Wang, S. (2021). Say no to the discrimination: Learning fair graph neural networks with limited sensitive attribute information. In L. Lewin-Eytan, D. Carmel, E. Yom-Tov,

E. Agichtein, & E. Gabrilovich (Eds.), *WSDM '21, The Fourteenth ACM International Conference on Web Search and Data Mining, Virtual Event, Israel, March 8–12, 2021* (pp. 680–688). ACM.

34. Dai, H., Kozareva, Z., Dai, B., Smola, A., & Song, L. (2018). Learning steady-states of iterative algorithms over graphs. In *ICML*.

35. Dai, H., Li, H., Tian, T., Huang, X., Wang, L., Zhu, J., & Song, L. (2018). Adversarial attack on graph structured data. In *Proceedings of the 35th International Conference on Machine Learning, ICML 2018, Stockholmsmässan, Stockholm, Sweden, July 10-15, 2018. Proceedings of Machine Learning Research*, PMLR (Vol. 80, pp. 1123–1132).

36. Defferrard, M., Bresson, X., & Vandergheynst, P. (2016). Convolutional neural networks on graphs with fast localized spectral filtering. In D. D. Lee, M. Sugiyama, U. von Luxburg, I. Guyon, & R. Garnett (Eds.), *Advances in Neural Information Processing Systems 29: Annual Conference on Neural Information Processing Systems 2016, December 5–10, 2016, Barcelona, Spain* (pp. 3837–3845).

37. Defferrard, M., Bresson, X., & Vandergheynst, P. (2016). Convolutional neural networks on graphs with fast localized spectral filtering. In *NeurIPS* (pp. 3844–3852).

38. Dempster, A. P., Laird, N. M., & Rubin, D. B. (1977). Maximum likelihood from incomplete data via the em algorithm. *Journal of the Royal Statistical Society: Series B (Methodological), 39*(1), 1–22.

39. Deng, X., & Zhang, Z. (2021). Graph-free knowledge distillation for graph neural networks. arXiv:2105.07519.

40. Devlin, J., Chang, M., Lee, K., & Toutanova, K. (2019). BERT: Pre-training of deep bidirectional transformers for language understanding. In *NAACL-HLT* (pp. 4171–4186).

41. Dhingra, B., Shallue, C. J., Norouzi, M., Dai, A. M., & Dahl, G. E. (2018). Embedding text in hyperbolic spaces. arXiv:1806.04313.

42. Dong, Y., Chawla, N. V., & Swami, A. (2017). Metapath2vec: Scalable representation learning for heterogeneous networks. In *SIGKDD* (pp. 135–144).

43. Du, M., Yang, F., Zou, N., & Hu, X. (2021). Fairness in deep learning: A computational perspective. *IEEE Intelligent Systems, 36*(4), 25–34.

44. Du, Y., Guo, X., Cao, H., Ye, Y., & Zhao, L. (2022). Disentangled spatiotemporal graph generative models. arXiv:2203.00411.

45. Fan, S., Zhu, J., Han, X., Shi, C., Hu, L., Ma, B., & Li, Y. (2019). Metapath-guided heterogeneous graph neural network for intent recommendation. In *KDD* (pp. 2478–2486).

46. Fan, W., Ma, Y., Li, Q., He, Y., Zhao, E., Tang, J., & Yin, D. (2019). Graph neural networks for social recommendation. In *The World Wide Web Conference* (pp. 417–426).

47. Fan, Y., Hou, S., Zhang, Y., Ye, Y., & Abdulhayoglu, M. (2018). Gotcha - sly malware!: Scorpion A metagraph2vec based malware detection system. In *KDD* (pp. 253–262).

48. Fang, G., Song, J., Shen, C., Wang, X., Chen, D., & Song, M. (2019). Data-free adversarial distillation. arXiv:1912.11006.

49. Fawcett, T. (2006). An introduction to roc analysis. *Pattern Recognition Letters, 27*(8), 861–874.

50. Fey, M., & Lenssen, J. E. (2019). Fast graph representation learning with pytorch geometric. arXiv:1903.02428.

51. Fout, A., Byrd, J., Shariat, B., & Ben-Hur, A. (2017). Protein interface prediction using graph convolutional networks. In *NIPS*.

52. Franceschi, L., Niepert, M., Pontil, M., & He, X. (2019). Learning discrete structures for graph neural networks. In *ICML* (Vol. 97, pp. 1972–1982).

53. Fréchet, M. (1948). Les éléments aléatoires de nature quelconque dans un espace distancié. In *Annales de l'institut Henri Poincaré* (Vol. 10, pp. 215–310).

54. Fu, T., Lee, W., & Lei, Z. (2017). Hin2vec: Explore meta-paths in heterogeneous information networks for representation learning. In *CIKM* (pp. 1797–1806).

55. Fu, X., Zhang, J., Meng, Z., & King, I. (2020). MAGNN: Metapath aggregated graph neural network for heterogeneous graph embedding. In Y. Huang, I. King, T. Liu, & M. van Steen (Eds.), *WWW '20: The Web Conference 2020, Taipei, Taiwan, April 20–24, 2020* (pp. 2331–2341). ACM/IW3C2.

56. Fu, X., Zhang, J., Meng, Z., & King, I. (2020). MAGNN: Metapath aggregated graph neural network for heterogeneous graph embedding. In *WWW* (pp. 2331–2341).

57. Furlanello, T., Lipton, Z., Tschannen, M., Itti, L., & Anandkumar, A. (2018). Born again neural networks. In *International Conference on Machine Learning* (pp. 1607–1616).

58. Ganea, O., Bécigneul, G., & Hofmann, T. (2018). Hyperbolic entailment cones for learning hierarchical embeddings. In *ICML* (pp. 1646–1655).

59. Ganea, O., Bécigneul, G., & Hofmann, T. (2018). Hyperbolic neural networks. In *NeurIPS* (pp. 5350–5360).

60. Gao, H., & Ji, S. (2019). Graph u-nets. In K. Chaudhuri & R. Salakhutdinov (Eds.), *Proceedings of the 36th International Conference on Machine Learning, ICML 2019, 9–15 June 2019, Long Beach, California, USA. Proceedings of Machine Learning Research* (Vol. 97, pp. 2083–2092), PMLR.

61. Gilmer, J., Schoenholz, S. S., Riley, P. F., Vinyals, O., & Dahl, G. E. (2017). Neural message passing for quantum chemistry. In *ICML* (Vol. 70, pp. 1263–1272).

62. Girvan, M., & Newman, M. E. (2002). Community structure in social and biological networks. *Proceedings of the National Academy of Sciences, 99*(12), 7821–7826.

63. Goodfellow, I., Pouget-Abadie, J., Mirza, M., Xu, B., Warde-Farley, D., Ozair, S., Courville, A., & Bengio, Y. (2014). Generative adversarial nets. *Advances in neural information processing systems 27*.

64. Goodfellow, I. J., Shlens, J., & Szegedy, C. (2015). Explaining and harnessing adversarial examples. In Y. Bengio & Y. LeCun (Eds.), *3rd International Conference on Learning Representations, ICLR 2015, San Diego, CA, USA, May 7–9, 2015, Conference Track Proceedings*.

65. Gori, M., Monfardini, G., & Scarselli, F. (2005). A new model for learning in graph domains. In *IJCNN* (pp. 729–734).

66. Gou, J., Yu, B., Maybank, S. J., & Tao, D. (2021). Knowledge distillation: A survey. *International Journal of Computer Vision, 129*(6), 1789–1819.

67. Goyal, P., Kamra, N., He, X., & Liu, Y. (2018). Dyngem: Deep embedding method for dynamic graphs. arXiv:1805.11273.

68. Gretton, A., Bousquet, O., Smola, A., & Schölkopf, B. (2005). Measuring statistical dependence with hilbert-schmidt norms. In *ALT* (pp. 63–77).

69. Gromov, M. (1987). Hyperbolic groups. In *Essays in group theory* (pp. 75–263). Springer.

70. Grover, A., & Leskovec, J. (2016). node2vec: Scalable feature learning for networks. In *Proceedings of the 22nd ACM SIGKDD International Conference on Knowledge Discovery and Data Mining* (pp. 855–864).

71. Gülçehre, Ç., Denil, M., Malinowski, M., Razavi, A., Pascanu, R., Hermann, K. M., Battaglia, P. W., Bapst, V., Raposo, D., Santoro, A., & de Freitas, N. (2019). Hyperbolic attention networks. In *7th International Conference on Learning Representations, ICLR 2019, New Orleans, LA, USA, May 6–9, 2019*. OpenReview.net.

72. Gulrajani, I., Ahmed, F., Arjovsky, M., Dumoulin, V., & Courville, A. (2017). Improved training of wasserstein gans. arXiv:1704.00028.

73. Guo, J., Li, J., Leng, D., & Pan, L. (2021). Heterogeneous graph based deep learning for biomedical network link prediction. arXiv:2102.01649.

74. Hamilton, W., Ying, Z., & Leskovec, J. (2017). Inductive representation learning on large graphs. In *Advances in Neural Information Processing Systems* (pp. 1024–1034).

75. Hammond, D. K., Vandergheynst, P., & Gribonval, R. (2011). Wavelets on graphs via spectral graph theory. *Applied and Computational Harmonic Analysis, 30*(2), 129–150.

76. Hao, B., Zhang, J., Yin, H., Li, C., & Chen, H. (2021). Pre-training graph neural networks for cold-start users and items representation. In *Proceedings of the 14th ACM International Conference on Web Search and Data Mining*.

77. Hassani, K., & Ahmadi, A. H. K. (2020). Contrastive multi-view representation learning on graphs. In *ICML* (pp. 4116–4126).

78. He, K., Fan, H., Wu, Y., Xie, S., & Girshick, R. B. (2020). Momentum contrast for unsupervised visual representation learning. In *CVPR* (pp. 9726–9735).

79. Hinton, G., Vinyals, O., & Dean, J. (2015). Distilling the knowledge in a neural network. arXiv:1503.02531.

80. Hochreiter, S., & Schmidhuber, J. (1997). Long short-term memory. *Neural Computation, 9*(8), 1735–1780.

81. Holland, P. W., Laskey, K. B., & Leinhardt, S. (1983). Stochastic blockmodels: First steps. *Social Networks, 5*(2), 109–137.

82. Hu, B., Fang, Y., & Shi, C. (2019). Adversarial learning on heterogeneous information networks. In *KDD* (pp. 120–129).

83. Hu, L., Yang, T., Shi, C., Ji, H., & Li, X. (2019). Heterogeneous graph attention networks for semi-supervised short text classification. In *EMNLP-IJCNLP* (pp. 4823–4832).

84. Hu, Z., Dong, Y., Wang, K., & Sun, Y. (2020). Heterogeneous graph transformer. In *WWW* (pp. 2704–2710).

85. Huang, H., Shi, R., Zhou, W., Wang, X., Jin, H., & Fu, X. (2021). Temporal heterogeneous information network embedding. In *IJCAI* (pp. 1470–1476).

86. Huang, H., Tang, J., Liu, L., Luo, J., & Fu, X. (2015). Triadic closure pattern analysis and prediction in social networks. *IEEE Transactions on Knowledge and Data Engineering, 27*(12), 3374–3389.

87. Huang, Z., & Mamoulis, N. (2017). Heterogeneous information network embedding for meta path based proximity. arXiv:1701.05291.

88. Huang, Z., Zheng, Y., Cheng, R., Sun, Y., Mamoulis, N., & Li, X. (2016). Meta structure: Computing relevance in large heterogeneous information networks. In *Proceedings of the 22nd ACM SIGKDD International Conference on Knowledge Discovery and Data Mining* (pp. 1595–1604).

89. Ji, H., Li, P., Shi, C., & Yang, C. (2021). Heterogeneous graph neural network with distance encoding. In *ICDM*.

90. Ji, H., Wang, X., Shi, C., Wang, B., & Yu, P. S. (2021). Heterogeneous graph propagation network. *IEEE Transactions on Knowledge and Data Engineering*.

91. Ji, Y., Fang, Y., & Shi, C. (2021). Dynamic meta-path guided temporal heterogeneous graph neural networks. In *KDD Workshop HENA*.

92. Ji, Y., Jia, T., Fang, Y., & Shi, C. (2021). Dynamic heterogeneous graph embedding via heterogeneous hawkes process. In *ECML-PKDD* (pp. 388–403). Springer.

93. Ji, Y., Yin, M., Yang, H., Zhou, J., Zheng, V. W., Shi, C., & Fang, Y. (2020). Accelerating large-scale heterogeneous interaction graph embedding learning via importance sampling. *ACM Transactions on Knowledge Discovery from Data, 15*(1), 1–23.

94. Jiang, B., Zhang, Z., Lin, D., Tang, J., & Luo, B. (2019). Semi-supervised learning with graph learning-convolutional networks. In *CVPR* (pp. 11313–11320).

95. Jin, W., Ma, Y., Liu, X., Tang, X., Wang, S., & Tang, J. (2020). Graph structure learning for robust graph neural networks. In *KDD* (pp. 66–74).

96. Jing, Y., Yang, Y., Wang, X., Song, M., & Tao, D. (2021). Amalgamating knowledge from heterogeneous graph neural networks. In *Proceedings of the IEEE/CVF Conference on Computer Vision and Pattern Recognition* (pp. 15709–15718).

97. Karcher, H. (2014). Riemannian center of mass and so called karcher mean. arXiv:1407.2087.

98. Karrer, B., & Newman, M. E. J. (2011). Stochastic blockmodels and community structure in networks. *Physical Review E, 83*(1), 16107.

99. Kazemi, S. M., Goel, R., Jain, K., Kobyzev, I., Sethi, A., Forsyth, P., & Poupart, P. Representation learning for dynamic graphs: A survey. *Journal of Machine Learning Research, 21*, 70:1–70:73.

100. Kim, T., Oh, J., Kim, N., Cho, S., & Yun, S.-Y. (2021). Comparing kullback-leibler divergence and mean squared error loss in knowledge distillation. arXiv:2105.08919.

101. Kingma, D. P., & Ba, J. (2014). Adam: A method for stochastic optimization. arXiv:1412.6980.

102. Kipf, T. N., & Welling, M. (2017). Semi-supervised classification with graph convolutional networks. In *ICLR*.

103. Klicpera, J., Bojchevski, A., & Günnemann, S. (2019). Predict then propagate: Graph neural networks meet personalized pagerank. In *ICLR*.

104. Klicpera, J., Weißenberger, S., & Günnemann, S. (2019). Diffusion improves graph learning. In *Advances in Neural Information Processing Systems* (pp. 13354–13366).

105. Krioukov, D., Papadopoulos, F., Kitsak, M., Vahdat, A., & Boguná, M. (2010). Hyperbolic geometry of complex networks. *Physical Review E, 82*(3), 036106.

106. Lai, C., Han, J., & Dong, H. (2021). Tensorlayer 3.0: A deep learning library compatible with multiple backends. In *2021 IEEE International Conference on Multimedia & Expo Workshops (ICMEW)*, IEEE (pp. 1–3).

107. Lai, Y., Hsu, C., Chen, W., Yeh, M., & Lin, S. (2017). PRUNE: Preserving proximity and global ranking for network embedding. In I. Guyon, U. von Luxburg, S. Bengio, H. M. Wallach, R. Fergus, S. V. N. Vishwanathan, & R. Garnett (Eds.), *Advances in Neural Information Processing Systems 30: Annual Conference on Neural Information Processing Systems 2017, December 4–9, 2017, Long Beach, CA, USA* (pp. 5257–5266).

108. Lan, Z., Chen, M., Goodman, S., Gimpel, K., Sharma, P., & Soricut, R. (2020). ALBERT: A lite BERT for self-supervised learning of language representations. In *ICLR*.

109. Law, M., Liao, R., Snell, J., & Zemel, R. (2019). Lorentzian distance learning for hyperbolic representations. In *ICML* (pp. 3672–3681).

110. Lee, S., Park, S., Kahng, M., & Lee, S.-G. (2013). Pathrank: Ranking nodes on a heterogeneous graph for flexible hybrid recommender systems. *Expert Systems with Applications, 40*(2), 684–697.

111. Leimeister, M., & Wilson, B. J. (2018). Skip-gram word embeddings in hyperbolic space. arXiv:1809.01498.

112. Leskovec, J., Kleinberg, J., & Faloutsos, C. (2005). Graphs over time: Densification laws, shrinking diameters and possible explanations. In *KDD*, ACM (pp. 177–187).

113. Li, J., Dani, H., Hu, X., Tang, J., Chang, Y., & Liu, H. (2017). Attributed network embedding for learning in a dynamic environment. In E. Lim, M. Winslett, M. Sanderson, A. W. Fu, J. Sun, J. S. Culpepper, E. Lo, J. C. Ho, D. Donato, R. Agrawal, Y. Zheng, C. Castillo, A. Sun, V. S. Tseng, & C. Li (Eds.), *Proceedings of the 2017 ACM on Conference on Information and Knowledge Management, CIKM 2017, Singapore, November 06–10, 2017*, ACM (pp. 387–396).

114. Li, M., Zhou, J., Hu, J., Fan, W., Zhang, Y., Gu, Y., & Karypis, G. (2021). Dgl-lifesci: An open-source toolkit for deep learning on graphs in life science. *ACS Omega*.

115. Li, P., Wang, Y., Wang, H., & Leskovec, J. (2020). Distance encoding - design provably more powerful graph neural networks for structural representation learning. In *NIPS*.

116. Li, Q., Han, Z., & Wu, X. (2018). Deeper insights into graph convolutional networks for semi-supervised learning. In *AAAI* (pp. 3538–3545).

117. Li, Q., Wu, X.-M., Liu, H., Zhang, X., & Guan, Z. (2019). Label efficient semi-supervised learning via graph filtering. In *Proceedings of the IEEE Conference on Computer Vision and Pattern Recognition* (pp. 9582–9591).
118. Li, X., Ding, D., Kao, B., Sun, Y., & Mamoulis, N. (2020). Leveraging meta-path contexts for classification in heterogeneous information networks. arXiv:2012.10024.
119. Liu, M., Gao, H., & Ji, S. (2020). Towards deeper graph neural networks. In *KDD* (pp. 338–348).
120. Liu, Q., Nickel, M., & Kiela, D. (2019). Hyperbolic graph neural networks. In *NeurIPS* (pp. 8228–8239).
121. Liu, Y., Pan, S., Jin, M., Zhou, C., Xia, F., & Yu, P. S. (2021). Graph self-supervised learning: A survey. arXiv:2103.00111.
122. Lopes, R. G., Fenu, S., & Starner, T. (2017). Data-free knowledge distillation for deep neural networks. arXiv:1710.07535.
123. Lu, Y., Jiang, X., Fang, Y., & Shi, C. (2021). Learning to pre-train graph neural networks. In *Proceedings of the AAAI Conference on Artificial Intelligence*.
124. Lu, Y., Wang, X., Shi, C., Yu, P. S., & Ye, Y. (2019). Temporal network embedding with micro- and macro-dynamics. In *CIKM* (pp. 469–478).
125. Ma, Y., Guo, Z., Ren, Z., Tang, J., & Yin, D. (2020). Streaming graph neural networks. In *SIGIR* (pp. 719–728).
126. Ma, Y., Wang, S., Aggarwal, C. C., & Tang, J. (2019). Graph convolutional networks with eigenpooling. In A. Teredesai, V. Kumar, Y. Li, R. Rosales, E. Terzi, & G. Karypis (Eds.), *Proceedings of the 25th ACM SIGKDD International Conference on Knowledge Discovery & Data Mining, KDD 2019, Anchorage, AK, USA, August 4–8, 2019*, ACM (pp. 723–731).
127. Manessi, F., Rozza, A., & Manzo, M. (2020). Dynamic graph convolutional networks. *Pattern Recognition, 97*.
128. McAuley, J. J., Targett, C., Shi, Q., & van den Hengel, A. (2015). Image-based recommendations on styles and substitutes. In *SIGIR* (pp. 43–52).
129. McLachlan, G. J., & Krishnan, T. (2007). *The EM algorithm and extensions* (Vol. 382). New York: Wiley.
130. Meng, Z., Liang, S., Bao, H., & Zhang, X. (2019). Co-embedding attributed networks. In *WSDM* (pp. 393–401).
131. Mernyei, P., & Cangea, C. (2020). Wiki-cs: A wikipedia-based benchmark for graph neural networks. arXiv:2007.02901.
132. Mikolov, T., Sutskever, I., Chen, K., Corrado, G. S., & Dean, J. (2013). Distributed representations of words and phrases and their compositionality. In *NeurIPS* (pp. 3111–3119).
133. Monti, F., Boscaini, D., Masci, J., Rodolà, E., Svoboda, J., & Bronstein, M. M. (2017). Geometric deep learning on graphs and manifolds using mixture model cnns. In *2017 IEEE Conference on Computer Vision and Pattern Recognition, CVPR 2017, Honolulu, HI, USA, July 21–26, 2017*, IEEE Computer Society (pp. 5425–5434).
134. Muscoloni, A., Thomas, J. M., Ciucci, S., Bianconi, G., & Cannistraci, C. V. (2017). Machine learning meets complex networks via coalescent embedding in the hyperbolic space. *Nature Communications, 8*(1), 1615.
135. Nayak, G. K., Mopuri, K. R., Shaj, V., Radhakrishnan, V. B., & Chakraborty, A. (2019). Zero-shot knowledge distillation in deep networks. In *International Conference on Machine Learning*, PMLR (pp. 4743–4751).
136. Newman, M. (2018). Network structure from rich but noisy data. *Nature Physics, 14*(6), 542–545.
137. Newman, M. E., Strogatz, S. H., & Watts, D. J. (2001). Random graphs with arbitrary degree distributions and their applications. *Physical Review E, 64*(2), 026118.
138. Newman, M. E. J. (2010). *Networks: An introduction*. Oxford: Oxford University Press.

139. Nguyen, G. H., Lee, J. B., Rossi, R. A., Ahmed, N. K., Koh, E., & Kim, S. (2018). Continuous-time dynamic network embeddings. In P. Champin, F. Gandon, M. Lalmas, & P. G. Ipeirotis (Eds.), *Companion of the The Web Conference 2018 on The Web Conference 2018, WWW 2018, Lyon , France, April 23–27, 2018*, ACM (pp. 969–976).

140. Nickel, M., & Kiela, D. (2017). Poincaré embeddings for learning hierarchical representations. In I. Guyon, U. von Luxburg, S. Bengio, H. M. Wallach, R. Fergus, S. V. N. Vishwanathan, & R. Garnett (Eds.), *Advances in Neural Information Processing Systems 30: Annual Conference on Neural Information Processing Systems 2017, December 4–9, 2017, Long Beach, CA, USA* (pp. 6338–6347).

141. Nickel, M., & Kiela, D. (2017). Poincaré embeddings for learning hierarchical representations. In *NeurIPS* (pp. 6338–6347).

142. Nickel, M., & Kiela, D. (2018). Learning continuous hierarchies in the lorentz model of hyperbolic geometry. In *ICML* (pp. 3779–3788).

143. Niepert, M., Ahmed, M., & Kutzkov, K. (2016). Learning convolutional neural networks for graphs. In *International Conference on Machine Learning*, PMLR (pp. 2014–2023).

144. Niu, D., Dy, J. G., & Jordan, M. I. (2010). Multiple non-redundant spectral clustering views. In *ICML* (pp. 831–838).

145. NT, H., & Maehara, T. (2019). Revisiting graph neural networks: All we have is low-pass filters. arXiv:1905.09550.

146. Pal, S., & Mitra, S. (1992). Multilayer perceptron, fuzzy sets, and classification. *IEEE Transactions on Neural Networks, 3*(5), 683–697.

147. Pareja, A., Domeniconi, G., Chen, J., Ma, T., Suzumura, T., Kanezashi, H., Kaler, T., Schardl, T., & Leiserson, C. (2020). Evolvegcn: Evolving graph convolutional networks for dynamic graphs. In *AAAI* (Vol. 34, pp. 5363–5370).

148. Parisot, S., Ktena, S. I., Ferrante, E., Lee, M., Moreno, R. G., Glocker, B., & Rueckert, D. (2017). Spectral graph convolutions for population-based disease prediction. In *MICCAI* (pp. 177–185).

149. Park, C., Kim, D., Han, J., & Yu, H. (2020). Unsupervised attributed multiplex network embedding. In *AAAI* (pp. 5371–5378).

150. Paszke, A., Gross, S., Massa, F., Lerer, A., Bradbury, J., Chanan, G., Killeen, T., Lin, Z., Gimelshein, N., Antiga, L., Desmaison, A., Köpf, A., Yang, E., DeVito, Z., Raison, M., Tejani, A., Chilamkurthy, S., Steiner, B., Fang, L., Bai, J., & Chintala, S. (2019). Pytorch: An imperative style, high-performance deep learning library. In *NeurIPS* (pp. 8024–8035).

151. Pei, H., Wei, B., Chang, K. C., Lei, Y., & Yang, B. (2020). Geom-gcn: Geometric graph convolutional networks. In *ICLR*. OpenReview.net.

152. Peng, H., Du, B., Liu, M., Liu, M., Ji, S., Wang, S., Zhang, X., & He, L. (2021). Dynamic graph convolutional network for long-term traffic flow prediction with reinforcement learning. *Information Sciences, 578*, 401–416.

153. Peng, W., Varanka, T., Mostafa, A., Shi, H., & Zhao, G. (2021). Hyperbolic deep neural networks: A survey. arXiv:2101.04562.

154. Perozzi, B., Al-Rfou, R., & Skiena, S. (2014). Deepwalk: Online learning of social representations. In *Proceedings of the 20th ACM SIGKDD International Conference on Knowledge Discovery and Data Mining* (pp. 701–710).

155. Qiu, J., Dong, Y., Ma, H., Li, J., Wang, K., & Tang, J. (2018). Network embedding as matrix factorization: Unifying deepwalk, line, pte, and node2vec. In *WSDM*, ACM (pp. 459–467).

156. Qu, M., Bengio, Y., & Tang, J. (2019). GMNN: Graph markov neural networks. In K. Chaudhuri & R. Salakhutdinov (Eds.), *Proceedings of the 36th International Conference on Machine Learning, ICML 2019, 9–15 June 2019, Long Beach, California, USA. Proceedings of Machine Learning Research*, PMLR (Vol. 97, pp. 5241–5250).

157. Ratcliffe, J. G., Axler, S., & Ribet, K. (1994). *Foundations of hyperbolic manifolds* (Vol. 3). Berlin: Springer.
158. Rhee, S., Seo, S., & Kim, S. (2017). Hybrid approach of relation network and localized graph convolutional filtering for breast cancer subtype classification. arXiv:1711.05859.
159. Ribeiro, L. F., Saverese, P. H., & Figueiredo, D. R. (2017). struc2vec: Learning node representations from structural identity. In *SIGKDD* (pp. 385–394).
160. Riolo, M. A., & Newman, M. (2020). Consistency of community structure in complex networks. *Physical Review E, 101*(5), 052306.
161. Sala, F., De Sa, C., Gu, A., & Ré, C. (2018). Representation tradeoffs for hyperbolic embeddings. In *ICML* (pp. 4457–4466).
162. Scarselli, F., Gori, M., Tsoi, A. C., Hagenbuchner, M., & Monfardini, G. (2009). The graph neural network model. *IEEE Transactions on Neural Networks, 20*(1), 61–80.
163. Sen, P., Namata, G., Bilgic, M., Getoor, L., Galligher, B., & Eliassi-Rad, T. (2008). Collective classification in network data. *AI Magazine, 29*(3), 93–93.
164. Shang, J., Qu, M., Liu, J., Kaplan, L. M., Han, J., & Peng, J. (2016). Meta-path guided embedding for similarity search in large-scale heterogeneous information networks. arXiv:1610.09769.
165. Shchur, O., Mumme, M., Bojchevski, A., & Günnemann, S. (2018). Pitfalls of graph neural network evaluation. arXiv:1811.05868.
166. Shi, C., Hu, B., Zhao, W. X., & Yu, P. S. (2019). Heterogeneous information network embedding for recommendation. *IEEE Transactions on Knowledge and Data Engineering, 31*(2), 357–370.
167. Shi, C., Li, Y., Zhang, J., Sun, Y., & Philip, S. Y. (2017). A survey of heterogeneous information network analysis. *IEEE Transactions on Knowledge and Data Engineering, 29*(1), 17–37.
168. Shi, Y., Huang, Z., Feng, S., & Sun, Y. (2020). Masked label prediction: Unified massage passing model for semi-supervised classification. arXiv:2009.03509.
169. Shlomi, J., Battaglia, P. W., & Vlimant, J. R. (2021). Graph neural networks in particle physics. *Machine Learning: Science and Technology, 2*, 21001.
170. Shuman, D. I., Narang, S. K., Frossard, P., Ortega, A., & Vandergheynst, P. (2013). The emerging field of signal processing on graphs: Extending high-dimensional data analysis to networks and other irregular domains. *IEEE Signal Processing Magazine, 30*(3), 83–98.
171. Simonovsky, M., & Komodakis, N. (2018). Graphvae: Towards generation of small graphs using variational autoencoders. In *International Conference on Artificial Neural Networks* (pp. 412–422). Springer.
172. Song, L., Smola, A., Gretton, A., Borgwardt, K. M., & Bedo, J. (2007). Supervised feature selection via dependence estimation. In *ICML* (pp. 823–830).
173. Song, W., Xiao, Z., Wang, Y., Charlin, L., Zhang, M., & Tang, J. (2019). Session-based social recommendation via dynamic graph attention networks. In *WSDM*, ACM (pp. 555–563).
174. Srivastava, N., Hinton, G. E., Krizhevsky, A., Sutskever, I., & Salakhutdinov, R. (2014). Dropout: a simple way to prevent neural networks from overfitting. *Journal of Machine Learning Research, 15*(1), 1929–1958.
175. Sun, K., Lin, Z., & Zhu, Z. (2020). Multi-stage self-supervised learning for graph convolutional networks on graphs with few labeled nodes. In *Thirty-Fourth AAAI Conference on Artificial Intelligence*.
176. Sun, K., Zhu, Z., & Lin, Z. (2021). Adagcn: Adaboosting graph convolutional networks into deep models. In *ICLR*. OpenReview.net.
177. Sun, L., Dou, Y., Yang, C., Wang, J., Yu, P. S., He, L., & Li, B. (2018). Adversarial attack and defense on graph data: A survey. arXiv:1812.10528.
178. Sun, L., Zhang, Z., Zhang, J., Wang, F., Peng, H., Su, S., & Yu, P. S. (2021). Hyperbolic variational graph neural network for modeling dynamic graphs. In *Proceedings of the AAAI Conference on Artificial Intelligence* (pp. 4375–4383).

179. Sun, Y., Han, J., Yan, X., Yu, P. S., & Wu, T. (2011). Pathsim: Meta path-based top-k similarity search in heterogeneous information networks. *Proceedings of the VLDB Endowment, 4*(11), 992–1003.

180. Suzuki, R., Takahama, R., & Onoda, S. (2019). Hyperbolic disk embeddings for directed acyclic graphs. In K. Chaudhuri & R. Salakhutdinov (Eds.), *Proceedings of the 36th International Conference on Machine Learning, ICML 2019, 9–15 June 2019, Long Beach, California, USA. Proceedings of Machine Learning Research*, PMLR, (Vol. 97, pp. 6066–6075).

181. Tang, J., Qu, M., Wang, M., Zhang, M., Yan, J., & Mei, Q. (2015). Line: Large-scale information network embedding. In *Proceedings of the 24th International Conference on World Wide Web* (pp. 1067–1077).

182. Tang, J., Sun, J., Wang, C., & Yang, Z. (2009). Social influence analysis in large-scale networks. In *KDD*, ACM (pp. 807–816).

183. Tay, Y., Tuan, L. A., & Hui, S. C. (2018). Hyperbolic representation learning for fast and efficient neural question answering. In *WSDM* (pp. 583–591).

184. Tian, Y., Krishnan, D., & Isola, P. (2020). Contrastive multiview coding. In *ECCV* (pp. 776–794).

185. Tifrea, A., Bécigneul, G., & Ganea, O.-E. (2018). Poincaré glove: Hyperbolic word embeddings. In *ICLR*.

186. Trivedi, R., Farajtabar, M., Biswal, P., & Zha, H. (2019). Dyrep: Learning representations over dynamic graphs. In *7th International Conference on Learning Representations, ICLR 2019, New Orleans, LA, USA, May 6–9, 2019*. OpenReview.net.

187. Vaswani, A., Shazeer, N., Parmar, N., Uszkoreit, J., Jones, L., Gomez, A. N., Kaiser, L., & Polosukhin, I. (2017). Attention is all you need. In I. Guyon, U. von Luxburg, S. Bengio, H. M. Wallach, R. Fergus, S. V. N. Vishwanathan, and R. Garnett (Eds.), *Advances in Neural Information Processing Systems 30: Annual Conference on Neural Information Processing Systems 2017, December 4–9, 2017, Long Beach, CA, USA* (pp. 5998–6008).

188. Velickovic, P., Fedus, W., Hamilton, W. L., Liò, P., Bengio, Y., & Hjelm, R. D. (2019). Deep graph infomax. In *ICLR*.

189. Veličković, P., Cucurull, G., Casanova, A., Romero, A., Liòò, P., & Bengio, Y. (2018). Graph attention networks. In *ICLR*.

190. Vinh, T. D. Q., Tay, Y., Zhang, S., Cong, G., & Li, X.-L. (2018). Hyperbolic recommender systems. arXiv:1809.01703.

191. Wang, D., Cui, P., & Zhu, W. (2016). Structural deep network embedding. In *Proceedings of the 22nd ACM SIGKDD international conference on Knowledge discovery and data mining* (pp. 1225–1234).

192. Wang, H., & Leskovec, J. (2020). Unifying graph convolutional neural networks and label propagation. arXiv:2002.06755.

193. Wang, M., Zheng, D., Ye, Z., Gan, Q., Li, M., Song, X., Zhou, J., Ma, C., Yu, L., Gai, Y., Xiao, T., He, T., Karypis, G., Li, J., & Zhang, Z. (2019). Deep graph library: A graph-centric, highly-performant package for graph neural networks. arXiv:1909.01315.

194. Wang, R., Mou, S., Wang, X., Xiao, W., Ju, Q., Shi, C., & Xie, X. (2021). Graph structure estimation neural networks. In *WWW*, ACM/IW3C2 (pp. 342–353).

195. Wang, R., Shi, C., Zhao, T., Wang, X., & Ye, Y. F. (2021). Heterogeneous information network embedding with adversarial disentangler. *IEEE Transactions on Knowledge and Data Engineering*.

196. Wang, W., Liu, X., Jiao, P., Chen, X., & Jin, D. (2018). A unified weakly supervised framework for community detection and semantic matching. In *PAKDD* (pp. 218–230).

197. Wang, X., Bo, D., Shi, C., Fan, S., Ye, Y., & Yu, P. S. (2020). A survey on heterogeneous graph embedding: Methods, techniques, applications and sources. arXiv:2011.14867.

198. Wang, X., Cui, P., Wang, J., Pei, J., Zhu, W., & Yang, S. (2017). Community preserving network embedding. In S. P. Singh & S. Markovitch (Eds.), *Proceedings of the Thirty-First AAAI Conference on Artificial Intelligence, February 4–9, 2017, San Francisco, California, USA*, AAAI Press (pp. 203–209).

199. Wang, X., He, X., Wang, M., Feng, F., & Chua, T.-S. (2019). Neural graph collaborative filtering. In *SIGIR* (pp. 165–174).

200. Wang, X., Ji, H., Shi, C., Wang, B., Ye, Y., Cui, P., & Yu, P. S. (2019). Heterogeneous graph attention network. In *WWW* (pp. 2022–2032).

201. Wang, X., Liu, N., Han, H., & Shi, C. (2021). Self-supervised heterogeneous graph neural network with co-contrastive learning. arXiv:2105.09111.

202. Wang, X., Lu, Y., Shi, C., Wang, R., Cui, P., & Mou, S. (2020). Dynamic heterogeneous information network embedding with meta-path based proximity. *IEEE Transactions on Knowledge and Data Engineering*.

203. Wang, X., Wang, R., Shi, C., Song, G., & Li, Q. (2020). Multi-component graph convolutional collaborative filtering. In *AAAI* (pp. 6267–6274).

204. Wang, X., Zhang, Y., & Shi, C. (2019). Hyperbolic heterogeneous information network embedding. In *AAAI* (pp. 5337–5344).

205. Wang, X., Zhu, M., Bo, D., Cui, P., Shi, C., & Pei, J. (2020). AM-GCN: Adaptive multi-channel graph convolutional networks. In *KDD*, ACM (pp. 1243–1253).

206. Wilson, R. C., Hancock, E. R., Pekalska, E., & Duin, R. P. (2014). Spherical and hyperbolic embeddings of data. *IEEE Transactions on Pattern Analysis and Machine Intelligence, 36*(11), 2255–2269.

207. Wu, F., Jr., A. H. S., Zhang, T., Fifty, C., Yu, T., & Weinberger, K. Q. (2019). Simplifying graph convolutional networks. In *ICML. Proceedings of Machine Learning Research*, PMLR (Vol. 97, pp. 6861–6871).

208. Wu, F., Souza, A., Zhang, T., Fifty, C., Yu, T., & Weinberger, K. (2019). Simplifying graph convolutional networks. In *International Conference on Machine Learning* (pp. 6861–6871).

209. Wu, W., Liu, H., Zhang, X., Liu, Y., & Zha, H. (2020). Modeling event propagation via graph biased temporal point process. *IEEE Transactions on Neural Networks and Learning Systems*.

210. Wu, Z., Pan, S., Chen, F., Long, G., Zhang, C., & Philip, S. Y. (2020). A comprehensive survey on graph neural networks. *IEEE Transactions on Neural Networks and Learning Systems*.

211. Xie, Y., Xu, Z., Wang, Z., & Ji, S. (2021). Self-supervised learning of graph neural networks: A unified review. arXiv:2102.10757.

212. Xu, B., Shen, H., Cao, Q., Cen, K., & Cheng, X. (2019). Graph convolutional networks using heat kernel for semi-supervised learning. In *IJCAI* (pp. 1928–1934). ijcai.org.

213. Xu, B., Shen, H., Cao, Q., Qiu, Y., & Cheng, X. (2019). Graph wavelet neural network. In *7th International Conference on Learning Representations, ICLR 2019, New Orleans, LA, USA, May 6–9, 2019*. OpenReview.net.

214. Xu, D., Ruan, C., Körpeoglu, E., Kumar, S., & Achan, K. (2020). Inductive representation learning on temporal graphs. In *8th International Conference on Learning Representations, ICLR 2020, Addis Ababa, Ethiopia, April 26–30, 2020*. OpenReview.net.

215. Xu, K., Hu, W., Leskovec, J., & Jegelka, S. (2019). How powerful are graph neural networks? In *ICLR*. OpenReview.net.

216. Xu, K., Li, C., Tian, Y., Sonobe, T., Kawarabayashi, K.-i., & Jegelka, S. (2018). Representation learning on graphs with jumping knowledge networks. In *ICML* (pp. 5453–5462).

217. Xue, H., Yang, L., Jiang, W., Wei, Y., Hu, Y., & Lin, Y. (2020). Modeling dynamic heterogeneous network for link prediction using hierarchical attention with temporal rnn. arXiv:2004.01024.

218. Yan, B., Wang, C., Guo, G., & Lou, Y. (2020). Tinygnn: Learning efficient graph neural networks. In *Proceedings of the 26th ACM SIGKDD International Conference on Knowledge Discovery & Data Mining* (pp. 1848–1856).

219. Yang, C., Liu, J., & Shi, C. (2021). Extract the knowledge of graph neural networks and go beyond it: An effective knowledge distillation framework. In *Proceedings of the Web Conference* (pp. 1227–1237).

220. Yang, Y., Qiu, J., Song, M., Tao, D., & Wang, X. (2020). Distilling knowledge from graph convolutional networks. In *Proceedings of the IEEE/CVF Conference on Computer Vision and Pattern Recognition* (pp. 7074–7083).

221. Yang, Z., Cohen, W. W., & Salakhutdinov, R. (2016). Revisiting semi-supervised learning with graph embeddings. arXiv:1603.08861.

222. Yin, Y., Ji, L.-X., Zhang, J.-P., & Pei, Y.-L. (2019). Dhne: Network representation learning method for dynamic heterogeneous networks. *IEEE Access, 7*, 134782–134792.

223. Ying, R., He, R., Chen, K., Eksombatchai, P., Hamilton, W. L., & Leskovec, J. (2018). Graph convolutional neural networks for web-scale recommender systems. In *SIGKDD* (pp. 974–983).

224. Ying, Z., Bourgeois, D., You, J., Zitnik, M., & Leskovec, J. (2019). Gnnexplainer: Generating explanations for graph neural networks. In H. M. Wallach, H. Larochelle, A. Beygelzimer, F. d'Alché-Buc, E. B. Fox, & R. Garnett (Eds.), *Advances in Neural Information Processing Systems 32: Annual Conference on Neural Information Processing Systems 2019, NeurIPS 2019, December 8-14, 2019, Vancouver, BC, Canada* (pp. 9240–9251).

225. Ying, Z., You, J., Morris, C., Ren, X., Hamilton, W. L., & Leskovec, J. (2018). Hierarchical graph representation learning with differentiable pooling. In S. Bengio, H. M. Wallach, H. Larochelle, K. Grauman, N. Cesa-Bianchi, and R. Garnett (Eds.), *Advances in Neural Information Processing Systems 31: Annual Conference on Neural Information Processing Systems 2018, NeurIPS 2018, December 3–8, 2018, Montréal, Canada* (pp. 4805–4815).

226. You, J., Ying, R., & Leskovec, J. (2019). Position-aware graph neural networks. In *ICML* (pp. 7134–7143).

227. You, J., Ying, R., Ren, X., Hamilton, W. L., & Leskovec, J. (2018). Graphrnn: Generating realistic graphs with deep auto-regressive models. In *ICML* (Vol. 80, pp. 5694–5703).

228. You, J., Ying, Z., & Leskovec, J. (2020). Design space for graph neural networks. *Advances in Neural Information Processing Systems, 33*, 17009–17021.

229. Yuan, H., Yu, H., Gui, S., & Ji, S. (2020). Explainability in graph neural networks: A taxonomic survey. arXiv:2012.15445.

230. Yun, S., Jeong, M., Kim, R., Kang, J., & Kim, H. J. (2019). Graph transformer networks. In *NeurIPS* (pp. 11960–11970).

231. Zang, C., Cui, P., & Faloutsos, C. (2016). Beyond sigmoids: The nettide model for social network growth, and its applications. In *KDD*, ACM (pp. 2015–2024).

232. Zarrinkalam, F., Kahani, M., & Bagheri, E. (2018). Mining user interests over active topics on social networks. *Information Processing & Management, 54*(2), 339–357.

233. Zhang, C., Song, D., Huang, C., Swami, A., & Chawla, N. V. (2019). Heterogeneous graph neural network. In *KDD* (pp. 793–803).

234. Zhang, M., & Chen, Y. (2018). Link prediction based on graph neural networks. In *NIPS* (pp. 5165–5175).

235. Zhang, M., Cui, Z., Neumann, M., & Chen, Y. (2018). An end-to-end deep learning architecture for graph classification. In *Thirty-Second AAAI Conference on Artificial Intelligence*.

236. Zhang, M., Hu, L., Shi, C., & Wang, X. (2020). Adversarial label-flipping attack and defense for graph neural networks. In C. Plant, H. Wang, A. Cuzzocrea, C. Zaniolo, & X. Wu (Eds.), *20th IEEE International Conference on Data Mining, ICDM 2020, Sorrento, Italy, November 17–20, 2020*, IEEE (pp. 791–800).

237. Zhang, M., Wu, S., Yu, X., Liu, Q., & Wang, L. (2021). Dynamic graph neural networks for sequential recommendation. arXiv:2104.07368.
238. Zhang, S., Liu, Y., Sun, Y., & Shah, N. (2021). Graph-less neural networks: Teaching old mlps new tricks via distillation. arXiv:2110.08727.
239. Zhang, W., Jiang, Y., Li, Y., Sheng, Z., Shen, Y., Miao, X., Wang, L., Yang, Z., & Cui, B. (2021). Rod: Reception-aware online distillation for sparse graphs. In *Proceedings of the 27th ACM SIGKDD Conference on Knowledge Discovery & Data Mining* (pp. 2232–2242).
240. Zhang, W., Miao, X., Shao, Y., Jiang, J., Chen, L., Ruas, O., & Cui, B. (2020). Reliable data distillation on graph convolutional network. In *Proceedings of the 2020 ACM SIGMOD International Conference on Management of Data* (pp. 1399–1414).
241. Zhang, Y., Pal, S., Coates, M., & Üstebay, D. (2019). Bayesian graph convolutional neural networks for semi-supervised classification. In *AAAI* (pp. 5829–5836).
242. Zhang, Y., Wang, X., Shi, C., Jiang, X., & Ye, Y. F. (2021). Hyperbolic graph attention network. *IEEE Transactions on Big Data*.
243. Zhang, Y., Wang, X., Shi, C., Liu, N., & Song, G. (2021). Lorentzian graph convolutional networks. In *WWW* (pp. 1249–1261).
244. Zhang, Y., Xiong, Y., Kong, X., Li, S., Mi, J., & Zhu, Y. (2018). Deep collective classification in heterogeneous information networks. In P. Champin, F. Gandon, M. Lalmas, & P. G. Ipeirotis (Eds.), *Proceedings of the 2018 World Wide Web Conference on World Wide Web, WWW 2018, Lyon, France, April 23–27, 2018*, ACM (pp. 399–408).
245. Zhang, Z., Cui, P., & Zhu, W. (2020). Deep learning on graphs: A survey. *IEEE Transactions on Knowledge and Data Engineering*.
246. Zhang, Z., & Sabuncu, M. (2020). Self-distillation as instance-specific label smoothing. *Advances in Neural Information Processing Systems 33*.
247. Zhao, J., Wang, X., Shi, C., Hu, B., Song, G., & Ye, Y. (2021). Heterogeneous graph structure learning for graph neural networks. In *AAAI*.
248. Zhao, J., Wang, X., Shi, C., Liu, Z., & Ye, Y. (2020). Network schema preserving heterogeneous information network embedding. In *IJCAI* (pp. 1366–1372).
249. Zhao, J., Zhou, Z., Guan, Z., Zhao, W., Ning, W., Qiu, G., & He, X. (2019). Intentgc: A scalable graph convolution framework fusing heterogeneous information for recommendation. In *KDD* (pp. 2347–2357).
250. Zheng, D., Song, X., Ma, C., Tan, Z., Ye, Z., Dong, J., Xiong, H., Zhang, Z., & Karypis, G. (2020). Dgl-ke: Training knowledge graph embeddings at scale. In *Proceedings of the 43rd International ACM SIGIR Conference on Research and Development in Information Retrieval* (New York, NY, USA, 2020), SIGIR '20, Association for Computing Machinery (pp. 739–748).
251. Zheng, J., Li, Q., & Liao, J. (2021). Heterogeneous type-specific entity representation learning for recommendations in e-commerce network. *Informantion Processing & Management, 58*(5), 102629.
252. Zheng, V. W., Sha, M., Li, Y., Yang, H., Fang, Y., Zhang, Z., Tan, K., & Chang, K. C. (2018). Heterogeneous embedding propagation for large-scale e-commerce user alignment. In *ICDM* (pp. 1434–1439).
253. Zheng, Y., Zhang, X., Chen, S., Zhang, X., Yang, X., & Wang, D. (2021). When convolutional network meets temporal heterogeneous graphs: An effective community detection method. *IEEE Transactions on Knowledge and Data Engineering*.
254. Zhou, L., Yang, Y., Ren, X., Wu, F., & Zhuang, Y. (2018). Dynamic network embedding by modeling triadic closure process. In *AAAI*.
255. Zhu, L., Guo, D., Yin, J., Ver Steeg, G., & Galstyan, A. (2016). Scalable temporal latent space inference for link prediction in dynamic social networks. *IEEE Transactions on Knowledge and Data Engineering, 28*(10), 2765–2777.

256. Zhu, S., Pan, S., Zhou, C., Wu, J., Cao, Y., & Wang, B. (2020). Graph geometry interaction learning. *NeurIPS,* 633–643.
257. Zhu, X., & Ghahramani, Z. (2002). Learning from labeled and unlabeled data with label propagation. *Technical Report CMU-CALD-02-107, Carnegie Mellon University.*
258. Zill, D., Wright, W. S., & Cullen, M. R. (2011). *Advanced engineering mathematics.* Jones & Bartlett Learning.
259. Zügner, D., Akbarnejad, A., & Günnemann, S. (2018). Adversarial attacks on neural networks for graph data. In *Proceedings of the 24th ACM SIGKDD International Conference on Knowledge Discovery & Data Mining, KDD 2018, London, UK, August 19–23, 2018,* ACM (pp. 2847–2856).
260. Zuo, Y., Liu, G., Lin, H., Guo, J., Hu, X., & Wu, J. (2018). Embedding temporal network via neighborhood formation. In *KDD,* ACM (pp. 2857–2866).

Printed in the United States
by Baker & Taylor Publisher Services